Advances in Analytics and Data Science

Volume 2

Series Editors
Ramesh Sharda
Oklahoma State University, Stillwater, OK, USA

Hsinchun Chen
University of Arizona, Tucson, AZ, USA

More information about this series at http://www.springer.com/series/15876

Murugan Anandarajan • Chelsey Hill
Thomas Nolan

Practical Text Analytics

Maximizing the Value of Text Data

Murugan Anandarajan
LeBow College of Business
Drexel University
Philadelphia, PA, USA

Chelsey Hill
Feliciano School of Business
Montclair State University
Montclair, NJ, USA

Thomas Nolan
Mercury Data Science
Houston, TX, USA

ISSN 2522-0233   ISSN 2522-0241   (electronic)
Advances in Analytics and Data Science
ISBN 978-3-319-95662-6   ISBN 978-3-319-95663-3   (eBook)
https://doi.org/10.1007/978-3-319-95663-3

Library of Congress Control Number: 2018955905

This Springer imprint is published by the registered company Springer Nature Switzerland AG
The registered company address is: Gewerbestrasse 11, 6330 Cham, Switzerland

To my aunts and uncle—MA
To my angel mother, Deborah—CH
To my dad—TN

Preface

The oft-cited statistic that "80% of data is unstructured" reminds industry leaders and analytics professionals about the vast volume of untapped text data resources. Predictably, there has been an increasing focus on text analytics to generate information resources. In fact, the expected growth in this market is projected to be $5.93 billion by 2020![1]

Whenever businesses capture text data, they want to capitalize on the information hidden within. Beginning with early adopters in the intelligence and biomedical communities, text analytics has expanded to include applications across industries, including manufacturing, insurance, healthcare, education, safety and security, publishing, telecommunications, and politics. The broad range of applied text analytics requires practitioners in this field.

Our goal is to democratize text analytics and increase the number of people using text data for research. We hope this book lowers the barrier of entry for analyzing text data, making it more accessible for people to uncover value-added text information.

This book covers the elements involved in creating a text mining pipeline. While analysts will not use every element in every project, each tool provides a potential segment in the final pipeline. Understanding the options is key to choosing the appropriate elements in designing and conducting text analysis.

The book is divided into five parts. The first part provides an overview of the text analytics process by introducing text analytics, discussing the relationship with content analysis, and providing a general overview of the process.

Next, the chapter moves on to the actual practice of text analytics, beginning with planning the project. The next part covers the methods of data preparation and preprocessing. Once the data is prepared, the next step is the analysis. Here, we describe the array of analysis options. The part concludes with a discussion about reporting options, indicating the benefits of various choices for convincing others about the value of the analysis.

---

[1] http://www.marketsandmarkets.com/PressReleases/text-analytics.asp

The last part of the book demonstrates the use of various software programs and programming languages for text analytics. We hope these examples provide the reader with practical examples of how information hidden within text data can be mined.

Philadelphia, PA, USA   Murugan Anandarajan
Montclair, NJ, USA   Chelsey Hill
Houston, TX, USA   Thomas Nolan

Acknowledgments

The authors wish to acknowledge the invaluable contributions of several individuals to the preparation of the manuscript of this book.

Diana Jones, Director of the Center for Business Analytics and the Dornsife Office for Experiential Learning, at the LeBow College of Business, Drexel University, for her chapter on *Storytelling Using Text Data*.

Jorge Fresneda Fernandez, Assistant Professor of Marketing at the Martin Tuchman School of Management, New Jersey Institute of Technology, for his chapters on *Latent Semantic Analysis (LSA) in Python* and *SAS Visual Text Analytics*.

We thank Diana and Jorge for their expertise and invaluable contributions to this book.

We also thank Irena Nedelcu, Rajiv Nag, and Stacy Boyer, all of the LeBow College of Business, Drexel University, for providing valuable comments on various chapters.

Our appreciation to Matthew Amboy and his team at Springer who made the publication of this book possible.

Murugan Anandarajan
Chelsey Hill
Thomas Nolan

Contents

About the Authors

Murugan Anandarajan is a Professor of MIS at Drexel University. His current research interests lie in the intersections of crime, IoT, and analytics. His work has been published in journals such as *Decision Sciences*, *Journal of Management Information Systems*, and *Journal of International Business Studies*. He co-authored eight books, including *The Internet and Workplace Transformation* (2006) and its follow-up volume, *The Internet of People, Things and Services* (2018). He has been awarded over $2.5 million in research grants from various government agencies including the National Science Foundation, the US Department of Justice, the National Institute of Justice, and the State of PA.

Chelsey Hill is an Assistant Professor of Business Analytics in the Information Management and Business Analytics Department of the Feliciano School of Business at Montclair State University. She holds a BA in Political Science from the College of New Jersey, an MS in Business Intelligence from Saint Joseph's University, and a PhD in Business Administration with a concentration in Decision Sciences from Drexel University. Her research interests include consumer product recalls, online consumer reviews, safety and security, public policy, and humanitarian operations. Her research has been published in the *Journal of Informetrics* and the *International Journal of Business Intelligence Research*.

Tom Nolan completed his undergraduate work at Kenyon College. After Kenyon, he attended Drexel University where he graduated with an MS in Business Analytics. From there, he worked at Independence Blue Cross in Philadelphia, PA, and Anthem Inc. in Houston, TX. Currently, he works with all types of data as a Data Scientist for Mercury Data Science.

List of Abbreviations

ANN	Artificial neural networks
BOW	Bag-of-words
CA	Content analysis
CTM	Correlated topic model
df	Document frequency
DM	Data mining
DTM	Document-term matrix
HCA	Hierarchical cluster analysis
idf	Inverse document frequency
IoT	Internet of Things
KDD	Knowledge discovery in databases
KDT	Knowledge discovery in text
kMC	k-means clustering
kNN	k-nearest neighbors
LDA	Latent Dirichlet allocation
LSA	Latent semantic analysis
LSI	Latent semantic indexing
NB	Naive Bayes
NLP	Natural language processing
NN	Neural networks
OM	Opinion mining
pLSI	Probabilistic latent semantic indexing
RF	Random forest
SA	Sentiment analysis
sLDA	Supervised latent Dirichlet allocation
STM	Structural topic model
SVD	Singular value decomposition
SVM	Support vector machines
TA	Text analytics
TDM	Term-document matrix
tf	Term frequency
tfidf	Term frequency-inverse document frequency
TM	Text mining

List of Figures

List of Tables

Chapter 1
Introduction to Text Analytics

Abstract In this chapter we define text analytics, discuss its origins, cover its current usage, and show its value to businesses. The chapter describes examples of current text analytics uses to demonstrate the wide array of real-world impacts. Finally, we present a process road map as a guide to text analytics and to the book.

Keywords Text analytics · Text mining · Data mining · Content analysis

1.1 Introduction

Recent estimates maintain that 80% of all data is text data. A recent article in *USA Today* asserts that even the Internal Revenue Service is using text, in the form of US citizens' social media posts, to help them make auditing decisions.[1] The importance of text data has created a veritable industry comprised of firms dedicated solely to the storage, analysis, and extraction of text data. One such company, Crimson Hexagon, has created the world's largest database of text data from social media sites including one trillion public social media posts spanning more than a decade.[2]

1.2 Text Analytics: What Is It?

Hearst (1999a, b) defines text analytics, sometimes referred to as text mining or text data mining, as the automatic discovery of new, previously unknown, information from unstructured textual data. The terms text analytics and text mining are often used interchangeably. Text mining can also be described as the process of deriving

---

[1] Agency breaking law by mining social media. (2017, 12). USA Today, 146, 14–15.

[2] http://www.businessinsider.com/analytics-firm-crimson-hexagon-uses-social-media-to-predict-stock-movements-2017-4

© Springer Nature Switzerland AG 2019

M. Anandarajan et al., *Practical Text Analytics*, Advances in Analytics and Data Science 2, https://doi.org/10.1007/978-3-319-95663-3_1

high-quality information from text. This process involves three major tasks: information retrieval (gathering the relevant documents), information extraction (unearthing information of interest from these documents), and data mining (discovering new associations among the extracted pieces of information).

Text analytics has been influenced by many fields and has made significant contributions to many disciplines. Modern text analytics applications span many disciplines and objectives. In addition to having multidisciplinary origins, text analytics continues to have applications in and make advancements to many fields of study. It intersects many research fields, including:

- Library and information science
- Social sciences
- Computer science
- Databases
- Data mining
- Statistics
- Artificial intelligence
- Computational linguistics

Although many of these areas contributed to modern day text analytics, Hearst (1999a, b) posited that text mining came to fruition as an extension of data mining. The similarities and differences between text mining and data mining have been widely discussed and debated (Gupta and Lehal 2009; Hearst 2003). Data mining uses structured data, typically from databases, to uncover "patterns, associations, changes, anomalies, and significant structures" (Bose 2009, p. 156). The major difference between the two is the types of data that they use for analysis. Data mining uses structured data, found in most business databases, while text mining uses unstructured or semi-structured data from a variety of sources, including media, the web, and other electronic data sources. The two methods are similar because they (i) are equipped for handling large data sets; (ii) look for patterns, insight, and discovery; and (iii) apply similar or the same techniques. Additionally, text mining draws on techniques used in data mining for the analysis of the numeric representation of text data.

The complexities associated with the collection, preparation, and analysis of unstructured text data make text analytics a unique area of research and application. Unstructured data are particularly difficult for computers to process. The data itself cover a wide range of possibilities, each with its own challenges. Some examples of the sources of text data used in text mining are blogs, web pages, emails, social media, message board posts, newspaper articles, journal articles, survey text, interview transcripts, resumes, corporate reports and letters, insurance claims, customer complaint letters, patents, recorded phone calls, contracts, and technical documentation (Bose 2009; Dörre et al. 1999).

1.3   Origins and Timeline of Text Analytics

Text-based analysis has its roots in the fields of computer science and the social sciences as a means of converting qualitative data into quantitative data for analysis. The field of computer science is in large part responsible for the text analytics that we know today. In contrast, the social sciences built the foundation of the analysis of text as a means of understanding literature, discourse, documents, and surveys. Text analytics combines the computational and humanistic elements of both fields and uses technology to analyze unstructured data text data by "turning text into numbers." The text analytics process includes the structuring of input text, deriving patterns within the structured data, and evaluating and interpreting the output.

Figure 1.1 presents a timeline of text analytics by decade. In the 1960s, computational linguistics was developed to describe computer-aided natural language processing (Miner et al. 2012). Natural language processing techniques are outlined in Chaps. 5 and 6. During this decade, content analysis, the focus of Chap. 2, emerged in the social sciences as a means of analyzing a variety of content, including text and media (Krippendorff 2012).

In the late 1980s and early 1990s, latent semantic indexing, or latent semantic analysis, introduced in Chap. 6 arrived as a dimension reduction and latent factor identification method applied to text (Deerwester et al. 1990; Dumais et al. 1988). At this time, knowledge discovery in databases developed as a means of making sense of data (Fayyad et al. 1996; Frawley et al. 1992). Building on this advancement, Feldman and Dagan (1995) created a framework for text, known as knowledge discovery in texts to do the same with unstructured text data.

Data mining emerged in the 1990s as the analysis step in the knowledge discovery in databases process (Fayyad et al. 1996). In the 1990s, machine learning methods, covered in Chaps. 7 and 9, gained prominence in the analysis of text data (Sebastiani 2002). Around that time, text mining became a popular buzzword but lacked practitioners (Hearst 1999a, b). Nagarkar and Kumbhar (2015) reviewed text mining-related publications and citations from 1999 to 2013 and found that the number of publications consistently increased throughout this period.

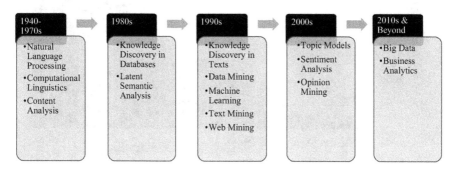

Fig. 1.1 Text analytics timeline

Building on a probabilistic form of latent semantic analysis (Hofmann 1999) introduced the late 1990s, topic models, discussed in Chap. 8, were created in the early 2000s with the development of the latent Dirichlet allocation model (Blei et al. 2002, 2003). Around the same time, sentiment analysis (Nasukawa and Yi 2003) and opinion mining (Dave et al. 2003), the focus of Chap. 10, were introduced as methods to understand and analyze opinions and feelings (Liu 2012).

The 2010s were the age of big data analytics. During this period, the foundational concepts preceding this time were adapted and applied to big data. According to Davenport (2013), although the field of business analytics has been around for over 50 years, the current era of analytics, Analytics 3.0, has witnessed the widespread use of corporate data for decision-making across many organizations and industries. More specifically, four key features define the current generation of text analytics and text mining: foundation, speed, logic, and output. As these characteristics indicate, text analysis and mining are data driven, conducted in real time, rely on probabilistic inference and models, and provide interpretable output and visualization (Müller et al. 2016). According to *IBM Tech Trends Report* (2011), business analytics was deemed one of the major trends in technology in the 2010s (Chen et al. 2012).

One way to better understand the area of text analytics is by examining relevant journal articles. We analyzed 3,264 articles, 2,315 published in scholarly journals and 949 published in trade journals. In our article sample, there are 704 journals, 187 trade journals and 517 scholarly journals. Each article belongs to one or more article classifications. The articles in our sample have 170 distinct classifications based on information about them such as industry, geographical location, content, and article type. Figure 1.2 displays the total number of text analytics-related

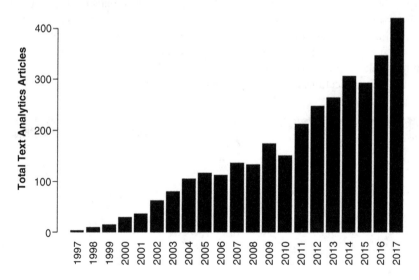

Fig. 1.2 Frequency of text analytics articles by year

articles over time. As the figure illustrates, there has been a considerable growth in the publications since 1997, with 420 articles about text mining published in 2017.

1.4 Text Analytics in Business and Industry

Text analytics has numerous practical uses, including but not limited to email filtering, product suggestions, fraud detection, opinion mining, trend analysis, search engines, and bankruptcy predictions (Talib et al. 2016). The field has a wide range of goals and objectives, including understanding semantic information, text summarization, classification, and clustering (Bolasco et al. 2005). We explore some examples of text analytics and text mining in the areas of business and industry. Text analytics applications require clear, interpretable results and actionable outcomes to achieve the desired result. Indeed, the technique can be used in almost every business department to increase productivity, efficiency, and understanding.

Figure 1.3 is a word cloud depicting the most popular terms and phrases in the text analytics-related articles' abstracts and titles. As the figure shows, research and applications in this area are as diverse as the area's history. The close relationship between data mining and text analytics is also evident in the figure.

Fig. 1.3 Word cloud of the titles and abstracts of articles on text analytics

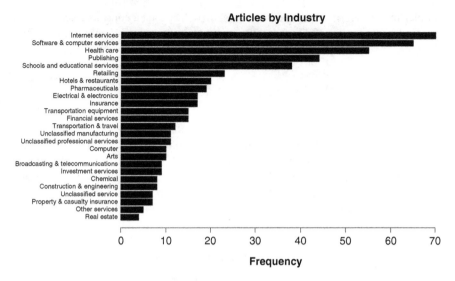

Fig. 1.4 Article frequency by industry for top 25 industry classifications

Applications of text analytics span a multitude of industries, including health-care, government, education, publishing, telecommunications, politics, and safety and security (Bolasco et al. 2005; Fan et al. 2006). Using our journal article sample, we analyze text analytics by industry. Figure 1.4 provides a plot of the frequency of the articles by industry. As the figure demonstrates, the publications cover many industries in which text analytics is a relevant area of research and practice, with manufacturing, services, computers, arts and broadcasting, and telecommunications being the most prevalent areas of research activity in the field.

1.5 Text Analytics Skills

Next, we consider the necessary skills for text mining applications. To do so, we collect and analyze a sample of 3,479 job postings listed on Indeed[3] using "text min-ing," "text analytics," and "text analysis" as our search keywords. By analyzing the descriptions of the jobs, their requirements, and their titles, we gain insight into the jobs that require text analysis skills and the specific skills, abilities, and experience that employers are seeking.

First, we analyze the content of the job titles of the postings to determine the types of jobs that require text analytics. As Fig. 1.5 shows, the most prevalent job title in which text analysis skills are required is data scientist. Text mining is also a

---

[3] https://www.indeed.com/

Fig. 1.5 Word cloud of text analytics job titles

Fig. 1.6 Word cloud of the skills required for text analytics jobs

desirable skill for data engineers, data and business analysts, marketing managers, and research analysts.

Figure 1.6 shows a word cloud created using the most popular terms and phrases describing the skills, abilities, and experience listed in the descriptions of the jobs. As the figure illustrates, NLP or natural language processing and machine learning modeling and software are the skills most sought after in text analysts. In this book, we introduce many of these topics, including text analysis methods, applications, and software.

1.6   Benefits of Text Analytics

The use of analytics in business can lead to many organization-wide benefits, including reductions in the time to realized value, organizational change, and stepwise achievement of goals (LaValle et al. 2011). Ur-Rahman and Harding (2012) suggest that the use of text mining in business can "cut overhead costs of product or service quality improvement and project management" (p. 4729). However, the reasons to conduct text analytics projects and the benefits of doing so are abundant.

First, such projects can help organizations make use of the estimated 80% of data that cannot be analyzed using data mining methods alone. Relying only on structured data can often mean missing part of the bigger picture. Second, text analytics can help companies index, catalogue, and store data for knowledge management and information retrieval. The use of text mining techniques can aid in knowledge management. For example, following a merger, Dow Union Carbide used text mining principles and methodology for knowledge management purposes, resulting in a substantial reduction in time, costs, and errors (Fan et al. 2006). Finally, text analytics can promote understanding in a climate of information overload. Text mining is often used for text summarization in cases where there is too much text for people to read. The potential efficiency and productivity gains from the application of text mining principles are undeniable.

1.7   Text Analytics Process Road Map

We break down the text analytics process into four major steps: (1) planning, (2) preparing and preprocessing, (3) analysis, and (4) reporting. Figure 1.7 depicts these four steps.

1.7.1   Planning

The planning stage determines the rest of the text analytics process, because the analyst sets the foundation for the analysis. In the planning chapters, we include an introduction to content analysis and a planning framework specific to text analytics applications. While the methodology for text analytics is well developed, the theoretical foundations are lacking. For this reason, we draw on the rich theory of content analysis applied to text analytics. To do so, we first present the fundamentals of content analysis in Chap. 2 and then develop a new planning process for text analytics projects in Chap. 3.

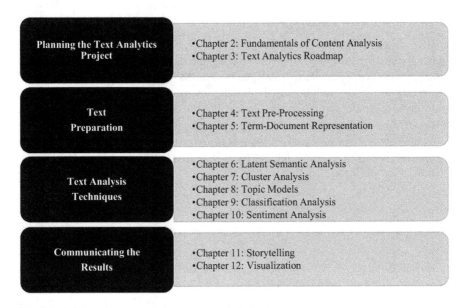

Fig. 1.7   Guide to the text analytics process and the book

1.7.2   Text Preparing and Preprocessing

In the preparing and preprocessing stage, the text data are prepared for analysis using appropriate software programs. In the area of data mining, this step is the point at which the data are cleansed and readied for analysis. In the context of text analytics, data reduction and preparation include many preprocessing tasks, which can be separated into two general steps: text preprocessing, introduced in Chap. 4, and term-document representation, outlined in Chap. 5.

1.7.3   Text Analysis Techniques

The chapters covering analysis in this book will help guide this decision-making process. There are two major types of analysis that we will present: unsupervised and supervised. In unsupervised analysis methods, no prior knowledge of the underlying document groupings is used in the analysis. In contrast, supervised analysis methods use known classifications, or categories, to build predictive models. We present three analysis methods that are predominantly unsupervised: latent semantic analysis in Chap. 6, cluster analysis in Chap. 7, and topic models in Chap. 8. We also review two methods that are typically supervised: classification analysis in Chap. 9 and sentiment analysis in Chap. 10.

1.7.4 Communicating the Results

After conducting the analysis, the final step is the interpretation of the findings and their significance. Regardless of the type of analysis used, reporting is a crucial step in sharing the results and making the findings actionable. The purpose of reporting is to provide details about the analysis and its results. Typically, reports include an overview of the problem and data along with more detailed information that is specific to the objectives and goals of the text analytics project. There are two key elements of text analytics reporting that we will explore: storytelling in Chap. 11 and visualization in Chap. 12.

1.8 Examples of Text Analytics Software

The choice of analysis tool can be very important to the project. For this reason, we include examples of applications in Chaps. 13, 14, 15 and 16 using real-world data in four software programs to complement the analysis chapters. The four software programs, depicted are R, Python, RapidMiner, and SAS.

Key Takeaways
- Text analytics is a diverse area with a rich interdisciplinary history.
- Text analytics has many applications in business and industry.
- The text analytics process outlined in this book has four components: planning preprocessing and preparing, analysis, and reporting.

References

Blei, D. M., Ng, A. Y., & Jordan, M. I. (2002). Latent Dirichlet allocation. In T. G. Dietterich, S. Becker, & Z. Ghahramani (Eds.), *Advances in neural information processing systems* (pp. 601–608). Cambridge: MIT Press.

Blei, D. M., Ng, A. Y., & Jordan, M. I. (2003). Latent Dirichlet allocation. *Journal of Machine Learning Research, 3*, 993–1022.

Bolasco, S., Canzonetti, A., Capo, F., Della Ratta-Rinaldi, F., & Singh, B. (2005). Understanding text mining: A pragmatic approach. In *Knowledge mining* (pp. 31–50). Heidelberg: Springer.

Bose, R. (2009). Advanced analytics: Opportunities and challenges. *Industrial Management & Data Systems, 109*(2), 155–172.

Chen, H., Chiang, R. H., & Storey, V. C. (2012). Business intelligence and analytics: From big data to big impact. *MIS Quarterly, 36*, 1165–1188.

Dave, K., Lawrence, S., & Pennock, D. M. (2003, May). Mining the peanut gallery: Opinion extraction and semantic classification of product reviews. In *Proceedings of the 12th International Conference on World Wide Web* (pp. 519–528). ACM.

Davenport, T. H. (2013). *Analytics 3.0*. Boston: Harvard Business Review.

Deerwester, S., Dumais, S. T., Furnas, G. W., Landauer, T. K., & Harshman, R. (1990). Indexing by latent semantic analysis. *Journal of the American Society for Information Science, 41*(6), 391.

Dörre, J., Gerstl, P., & Seiffert, R. (1999). Text mining: Finding nuggets in mountains of textual data. In *Proceedings of the fifth ACM SIGKDD International Conference on Knowledge Discovery and Data Mining* (pp. 398–401). ACM.

Dumais, S. T., Furnas, G. W., Landauer, T. K., & Deerwester, S. (1988). Using latent semantic analysis to improve information retrieval. In *Proceedings of CHI'88: Conference on Human Factors in Computing* (pp. 281–285). New York: ACM.

Fan, W., Wallace, L., Rich, S., & Zhang, Z. (2006). Tapping the power of text mining. *Communications of the ACM, 49*(9), 76–82. https://doi.org/10.1145/1151030.1151032.

Fayyad, U., Piatetsky-Shapiro, G., & Smyth, P. (1996). From data mining to knowledge discovery in databases. *AI Magazine, 17*(3), 37.

Feldman, R., & Dagan, I. (1995, August). Knowledge discovery in textual databases (KDT). *KDD, 95*, 112–117.

Frawley, W. J., Piatetsky-Shapiro, G., & Matheus, C. J. (1992). Knowledge discovery in databases: An overview. *AI Magazine, 13*(3), 57.

Gupta, V., & Lehal, G. S. (2009). A survey of text mining techniques and applications. *Journal of Emerging Technologies in Web Intelligence, 1*(1), 60–76.

Hearst, M. A. (1999a, June). Untangling text data mining. In *Proceedings of the 37th Annual Meeting of the Association for Computational Linguistics on Computational Linguistics* (pp. 3–10). Association for Computational Linguistics.

Hearst, M. A. (1999b). The use of categories and clusters for organizing retrieval results. In *Natural language information retrieval* (pp. 333–374). Dordrecht: Springer.

Hearst, M. (2003). *What is text mining*. UC Berkeley: SIMS.

Hofmann, T. (1999, July). Probabilistic latent semantic analysis. In *Proceedings of the Fifteenth Conference on Uncertainty in Artificial Intelligence* (pp. 289–296).

IBM. (2011, November 15). *The 2011 IBM tech trends report: The clouds are rolling in....is your business ready?* http://www.ibm.com/developerworks/techtrendsreport

Krippendorff, K. (2012). *Content analysis: An introduction to its methodology*. Thousand Oaks: Sage.

LaValle, S., Lesser, E., Shockley, R., Hopkins, M. S., & Kruschwitz, N. (2011). Big data, analytics and the path from insights to value. *MIT Sloan Management Review, 52*(2), 21–32.

Liu, B. (2012). Sentiment analysis and opinion mining. *Synthesis Lectures on Human Language Technologies, 5*(1), 1–167.

Miner, G., et al. (2012). *Practical text mining and statistical analysis for non-structured text data applications*. Amsterdam: Academic Press.

Müller, O., Junglas, I., Debortoli, S., & vom Brocke, J. (2016). Using text analytics to derive customer service management benefits from unstructured data. *MIS Quarterly Executive, 15*(4), 243–258.

Nagarkar, S. P., & Kumbhar, R. (2015). Text mining: An analysis of research published under the subject category 'information science library science' in web of science database during 1999–2013. *Library Review, 64*(3), 248–262.

Nasukawa, T., & Yi, J. (2003, October). Sentiment analysis: Capturing favorability using natural language processing. In *Proceedings of the 2nd International Conference on Knowledge Capture* (pp. 70–77). ACM.

Sebastiani, F. (2002). Machine learning in automated text categorization. *ACM Computing Surveys (CSUR), 34*(1), 1–47.

Talib, R., Hanif, M. K., Ayesha, S., & Fatima, F. (2016). Text mining: Techniques, applications and issues. *International Journal of Advanced Computer Science & Applications, 1*(7), 414–418.

Ur-Rahman, N., & Harding, J. A. (2012). Textual data mining for industrial knowledge management and text classification: A business oriented approach. *Expert Systems with Applications, 39*(5), 4729–4739.

Part I
Planning the Text Analytics Project

Chapter 2
The Fundamentals of Content Analysis

Abstract In this chapter, the reader is provided with an introduction to content analysis, which highlights the congruencies between content analysis and text analytics. The reader learns the differences between content types and is provided with a demonstration of the content analysis process. The chapter concludes with a discussion on how to properly manage the subject area's current theory for desired results.

Keywords Content analysis · Text analysis · Manifest content · Latent content · Inference · Deductive inference · Inductive inference · Unitizing · Coding · Analysis unit · Quantitative content analysis · Qualitative content analysis · Computer-aided text analysis

2.1 Introduction

In this book, we adopt a content analysis perspective due to its rich theoretical foundations. Text mining and analytics encompass a wide range of methods and techniques, but do not have a single, unifying foundation that spans the various areas of research and practice in the field. This book aims to present a guide to text mining within the framework of content analysis, to provide much-needed structure to the text analytics process.

Content analysis is "a research technique for making replicable and valid inferences from texts (or other meaningful matter) to the contexts of their use" (Krippendorff 2012, p. 18). Content analysis is specifically concerned with the analysis of qualitative data. Unlike quantitative, or numeric data, qualitative data are not innately measurable. For this reason, qualitative data require other techniques specifically tailored for their needs. Content analysis is one such technique for text analysis. We use the rigor of content analysis theory to guide our text analytics methodology.

M. Anandarajan et al., *Practical Text Analytics*, Advances in Analytics and Data Science 2, https://doi.org/10.1007/978-3-319-95663-3_2

2.2 Deductive Versus Inductive Approaches

As a research method for making inferences, or drawing conclusions, content analysis can either take a deductive or inductive approach. The role of theory in the analysis informs the approach used. Figure 2.1 displays the framework for content analysis and distinguishes the path of the analysis based on the type of inference. As shown, the deductive approach is driven by theory. Hypotheses and theory determine the dictionaries, or word lists, and categories, which are counted through a process known as coding to transform the qualitative data into a quantitative form for analysis. The inductive approach takes an open coding approach, in which the analysis is not driven by theory but by the data. Predetermined dictionaries and categories are not used during coding in content analysis for inductive inference.

2.2.1 Content Analysis for Deductive Inference

If the purpose of the research project is to test or retest a theory in a new context, a deductive approach is appropriate. Content analysis for deductive inference is the more traditional form of content analysis in which observed or manifest content is analyzed (Potter and Levine-Donnerstein 1999). Manifest content is that which physically appears on the page, with no interpretation of meaning.

This analysis is directed using existing theory and literature. The published theory tells the researcher the information to expect in the collected data including the topic's focal points and item definitions (Potter and Levine-Donnerstein 1999). The first step is to examine the current theory available on the topic. Obviously, this requirement means that there must be published theory for deductive content analysis to be appropriate (Elo and Kyngäs 2008). The researcher aims to understand the current explanations surrounding the research topic. The theory points the researcher to the important concepts contained within the research topic (Potter and Levine-Donnerstein 1999). This information is used in the next step to construct a hypothesis.

Hypotheses are used to test whether the expectations of the current theory match the collected data's content (Elo and Kyngäs 2008). These tests then use statistical methods and metrics to compare empirical findings with established theory to assess congruency. Hypotheses are tested, and the outcomes validate or refute existing theory. Three important features of content analysis for deductive inference are shown in Fig. 2.2.

2.2.2 Content Analysis for Inductive Inference

The inductive approach begins by assessing the data to extract all possible codes. Like the scientific method, inductive content analysis begins with the collected data and attempts to create a generalizable theory (Elo and Kyngäs 2008; Potter and

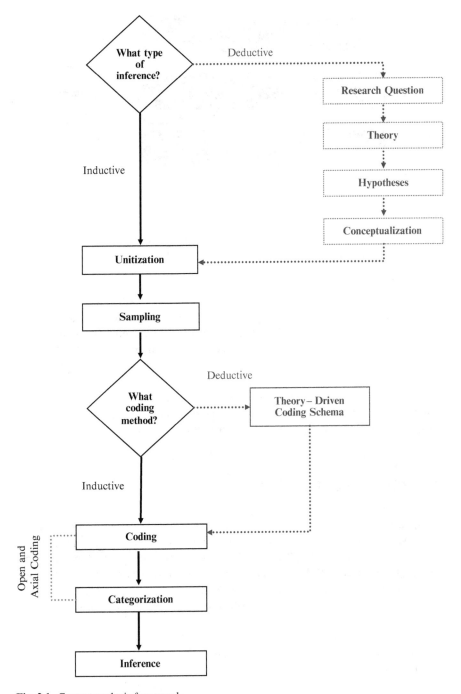

Fig. 2.1 Content analysis framework

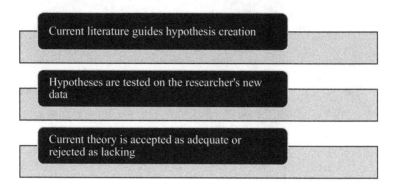

Fig. 2.2 Features of deductive inference

Fig. 2.3 Manifest and latent variables

Levine-Donnerstein 1999). This approach is appropriate when the researcher wants to build a new theory or expand on a current theory with the collected data.

Figure 2.3 visualizes the difference in the types of content that serve as the focus of the two types of content analysis. The figure describes the relationship between manifest and latent content. In a gray scale, more light reflects back, moving from black to white. Just as the black side of the gray scale is the absence of light, manifest content is devoid of interpretation. On the other hand, more of the coder's personal interpretations are reflected with latent content.

2.3 Unitizing and the Unit of Analysis

Next, the unit of analysis is chosen through a process known as unitization. The unit of analysis contains three parts: the sampling unit, the context unit, and the recording unit (Vourvachis 2007). The sampling unit is the text to be examined, corresponding to each row of data. In text analysis, the individual text items are known as documents. The recording unit is the text element to be categorized in the coding step. The context unit is the amount of text for determining the appropriate definition of the recording unit.

We use Preuss and Brown (2012) as an example to demonstrate the options for unitization. In this article, the authors content analyze the websites of companies listed on the Financial Times Stock Exchange 100 (FTSE 100) in 2009 to answer research questions about the frequency and content of human rights policies in large corporations.

2.3.1   The Sampling Unit

The sampling unit is the text used in the analysis, corresponding to each row of data or text document. The text can be anything, but there are some guidelines to follow when deciding what to use. First, the sampling unit must contain the information necessary to answer the research questions (Vourvachis 2007). An irrelevant sampling unit can lead to misleading or false inferences. Second, the size of the sampling unit should be considered. It needs to be large enough that the data are considered complete but limited to what is practical or feasible (Vourvachis 2007). In Sect. 2.4, and more comprehensively in Chap. 3, we present more about sampling.

In the sample article, the authors selected the firms' websites as the sampling unit. Each website was considered an individual text document in the sample used in the analysis. Corporate websites typically contain a large but manageable amount of information. We can assume that the authors knew that the corporate websites contained the appropriate data to answer the research questions about corporate policies. Thus, the use of company websites was a good choice for this study.

2.3.2   The Recording Unit

The recording unit designates the piece of text that is categorized in the coding step, described in Sect. 2.5. According to Holsti (1969), the recording unit is "the specific segment of content that is characterized by placing it in a given category" (p. 116). The recording unit could be a word, sentence, page, or an item specific to the project. The choice of the sampling unit impacts the choice of recording unit. At its largest, the recording unit can be the same size as the sampling unit (Vourvachis 2007). Words, being the smallest measurement option, provide the highest level of replicability. The disadvantage of words is that they can miss information, because they do not include context. This issue demonstrates the trade-offs the analyst must consider when selecting the recording unit. Smaller recording units are typically more reproducible, whereas larger ones incorporate contextual information but sacrifice reproducibility.

In our example, the authors used a combination of recording units to answer the various research questions. The first research question asked whether a company in the sample had a human rights policy. The researchers answered the first research question by using the corporate website as the recording unit. If the website included a human rights policy, the document for that company was coded as yes with respect to that research question. This case provides a good example of the research objective driving the decision-making.

2.3.3   The Context Unit

The context unit delineates the boundaries for questions about the meaning of the recording unit. It limits the amount of text used to place the recording unit in context. If the recording unit is a word and the context unit is a sentence, the sentence is used to help define the word. The context unit provides a path to define the recording unit when the meaning is ambiguous. It dictates the amount of text that can be utilized.

In the example, the researchers used the code of conduct and supporting documents as the context unit to address the research question regarding human rights. Drawing on prior research establishing the appropriateness of corporate codes of conduct as a context unit, the researchers chose these codes as the context unit.

2.4   Sampling

Sampling involves choosing a representative subset of a population. A population in content analysis includes all text documents that are relevant to answering the research question. In Preuss and Brown (2012), the population and sample were the same, and the corporate websites of all of the companies listed on the FTSE 100 were included in the analysis. While these researchers included the entire population in their sample, for larger populations, sampling is necessary to conduct an efficient and manageable analysis.

The sample is chosen from the relevant population of documents containing text (Stepchenkova 2012). A sample is considered representative of a population if the results of an analysis are approximately the same as what the results would be if the analysis were conducted on the entire population (Krippendorff 2012). For this reason, we need to consider the composition of the entire population. If the population is diverse, the sample should reflect that diversity. The main goal of sampling is to compile a group of documents from the larger population of documents that is representative of and relevant to the content analysis application. The sample should be created with the research objective in mind. In Chap. 3, we introduce sampling techniques used in text analytics.

2.5   Coding and Categorization

Coding refers to the classification of text data according to observer-independent rules (Krippendorff 2012, p. 126). Codes are used to create categories, which can be themes, topics, or classifications. Coding can be completed manually, as in human coding, or by computers, as in computer-aided coding. As shown in

Fig. 2.1, human coding is a more involved process than computer-based coding. In human coding, the coding scheme is first defined, because the coders must be trained to complete the coding properly (Neuendorf and Kumar 2006). After the trained coders complete the coding, inter-coder reliability is assessed. Reliability in content analysis coding is "the extent to which a measure in a coding scheme produces the same result when applied by different human coders" (Neuendorf and Kumar 2006, p. 3). In computer-aided coding, theoretically based dictionaries are used in coding. Even in computer-aided coding, there is still usually some need for human intervention and interpretation in most content analysis applications (Shapiro and Markoff 1997).

In addition to the two methods of coding, there are differences in the coding techniques based on the type of inference, inductive or deductive. Figure 2.4 gives a general overview of the difference. As shown, deductive inference is a top-down approach in which theory-driven categories predetermine the codes used, as in the case of dictionaries. In this approach, the text data are then coded within this theoretically driven framework, either by humans or computers. On the other hand, inductive inference takes a bottom-up approach in which the data influence the codes and the categories inform the resulting theory. This method is sometimes referred to as emergent coding.

The first step in coding for both inductive and deductive inference is open coding. In open coding, every theme is considered possible, and no interpretations of or connections between categories are made (Berg 1995). Strauss provides four strategies for effective open coding (Berg 1995; Strauss 1987), listed in Fig. 2.5.

The study's research question should be kept in mind and relevant questions are created. The data may not answer this question directly. That information should be recorded, too. Maintaining an open mind about additional information can result in interesting and unexpected findings. At first, the data should be analyzed minutely (Berg 1995). When beginning the coding, more is better. Therefore, it is done at a very detailed level to capture more information. Once code saturation has been

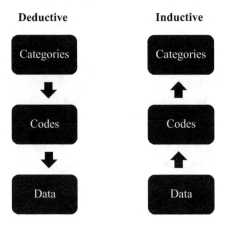

Fig. 2.4 Deductive and inductive coding approaches

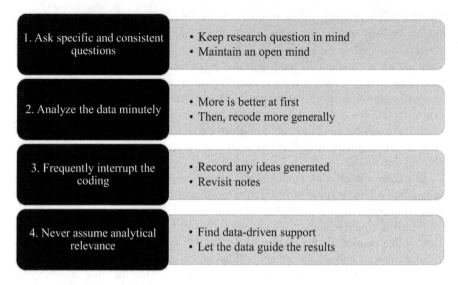

Fig. 2.5   Four open coding strategies

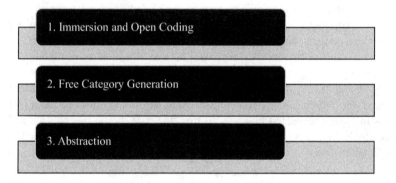

Fig. 2.6   The three-step inductive inference process

achieved, the coder steps back and codes more generally. Code saturation occurs when no new codes are being created.

Coding should be frequently interrupted so that theoretical notes can be written (Berg 1995). In the coding process, codes may spark theoretical ideas, which should be recorded in a separate file for future use. Without a written note to revisit, these ideas may be forgotten. Finally, the analyst should never assume the analytic relevance of any traditional variable until the data show it to be relevant or assume that descriptive variables (e.g., age, gender) have an impact without support from the data. Even if the research focuses on looking for differences in one of these variables, the collected data determine whether the variable is important (Berg 1995).

Once the initial codes have been created, the next step is to condense these codes into higher-level categories. This is an iterative process where each step aims to move from a specific to a more general category. The ultimate goal is to achieve generalizable and replicable categories that are applicable outside the collected sample of data. To achieve this outcome, we must condense the number of categories while maintaining the text's general information (Elo and Kyngäs 2008). The process takes the notes made during the open coding and places them under generated category headings.

Following the immersion and open coding step, coding for inductive analysis follows a more intensive than deductive process, as illustrated in Fig. 2.6. Having completed the open coding for inductive inference, in step 2, new categories can be generated freely. Since it is early in the process, it is better to create a category and combine it or delete it later than the reverse. After the initial iteration, appropriate categories can be further condensed. This process typically requires several iterations.

The final step in coding for inductive inference, the abstraction stage, produces the final categories. Their purpose is to provide a general response to the research question (Elo and Kyngäs 2008, p. 111). The categories should be at a high enough level that they are clear, concise, and relevant.

An alternative approach is axial coding. In axial coding, the category is the focus. In this method, the data are considered one category at a time. First, the initial category is chosen, and then all of the items that belong in that category are included. The process is repeated for each category necessary.

2.6 Examples of Inductive and Deductive Inference Processes

2.6.1 Inductive Inference

As an example of content analysis for inductive inference, we use Crane and Kazmi (2010), which considered the impact of business on children. In this study, the authors collected 357 accounts and labeled the important topics and themes. From this process, they extracted possible ideas for further analysis. They questioned each account's "[type] of impact (positive/negative), form of impact (direct or indirect), business sector, and relevant core business functions)" (Crane and Kazmi 2010, p. 6). The initial reading provided directions for condensing these ideas into interpretative codes. Content analysis allowed the authors to "identify the types of child impacts deemed relevant, newsworthy, and important by key stakeholders, rather than prescribing which issues 'should' be taken into consideration" (Crane and Kazmi 2010, p. 571). The authors used an inductive approach to produce categories directly from the collected data. Extant theory from outside the study did not influence their decisions about how to conduct the analysis. The seven categories created in Crane and Kazmi (2010) provided the researchers with descriptors of instances where "businesses were found to be confronted with issues of corporate responsibility toward young people" (p. 7). The final categories in their project were physical protection, moral protection, social and cultural participation, economic well-being, education and employability, parental and employment and family life, and impacting children's charities. These are good categories because they are helpful and generalizable and answer the research question.

2.6.2   Deductive Inference

We use Cho and Hambrick (2006) to illustrate deductive inference. This study investigated the organizational impact of the Airline Deregulation Act of 1978. Cho and Hambrick (2006) explored the conclusions drawn by industry observers, which the authors used as a guide. The publications contained what experts believed was happening to company attitudes in the industry. The published theory suggested that deregulation would result in companies shifting to an entrepreneurial focus—an important consequence of deregulation. From this knowledge, Cho and Hambrick (2006) constructed the hypothesis: "Following substantial deregulation, there will be a general shift in managerial attention toward more of an entrepreneurial orientation (relative to an engineering orientation)" (p. 4).

The authors tested six hypotheses in their study, but, for simplicity, only one is presented here. The authors content analyzed shareholder letters of publicly traded airline companies between 1973 and 1986. The first hypothesis sought to test a shift in an airline's entrepreneurial focus due to deregulation in the industry. Shareholder letters between 1973 and 1978 were written before the deregulation; the ones from 1979 to 1984 were published after the deregulation of the industry. The authors used the difference in frequency of theoretically defined words to test the hypothesis.

In the sample article, the publications regarding airline deregulation provide two sets of words. One set is commonly used in an entrepreneurial problem and the other in an engineering problem. The authors used these words to measure the degree to which an organization was focused on an entrepreneurial problem versus an engineering problem.

The decision about whether to use an inductive or deductive approach will impact the steps of the content analysis process and the interpretation of results. If theory is used as the foundation of the research, a top-down approach is undertaken. Conversely, without theory, a bottom-up approach is used, and the data dictate the outcome.

Key Takeaways
- Content analysis provides a robust methodological framework for the analysis of text.
- The chosen type of inference, deductive or inductive, guides the content analysis process.
- Four general steps in the content analysis process include unitization, sampling, coding, and inference.

References

Berg, B. L. (1995). *Qualitative research methods for the social sciences*. Boston: Bacon and Allyn.

Cho, T. S., & Hambrick, D. C. (2006). Attention as the mediator between top management team characteristics and strategic change: The case of airline deregulation. *Organization Science, 17*(4), 453–469.

Crane, A., & Kazmi, B. A. (2010). Business and children: Mapping impacts, managing responsibilities. *Journal of Business Ethics, 91*(4), 567–586.

Elo, S., & Kyngäs, H. (2008). The qualitative content analysis process. *Journal of Advanced Nursing, 62*(1), 107–115.

Holsti, O. R. (1969). *Content analysis for the social sciences and humanities*. Reading: Addison-Wesley Pub. Co.

Krippendorff, K. (2012). *Content analysis: An introduction to its methodology*. Los Angeles: Sage.

Neuendorf, K. A., & Kumar, A. (2002). Content analysis. *The International Encyclopedia of Political Communication, 1*, 221–230.

Potter, W. J., & Levine-Donnerstein, D. (1999). Rethinking validity and reliability in content analysis. *Journal of Applied Communications Research, 27*, 258–284.

Preuss, L., & Brown, D. (2012). Business policies on human rights: An analysis of their content and prevalence among FTSE 100 firms. *Journal of Business Ethics, 109*(3), 289–299.

Saldaña, J. (2015). *The coding manual for qualitative researchers*. Thousand Oaks: Sage.

Shapiro, G., & Markoff, J. (1997). A matter of definition. In C. W. Roberts (Ed.), *Text analysis for the social sciences: Methods for drawing statistical inferences from texts and transcripts*. Mahwah: Lawrence Erlbaum.

Stepchenkova, S. (2012). 23 content analysis. In *Handbook of research methods in tourism: Quantitative and qualitative approaches* (p. 443). Cheltenham: Edward Elgar.

Strauss, A. L. (1987). *Qualitative analysis for social scientists*. Cambridge: Cambridge University Press.

Vourvachis, P. (2007). On the use of content analysis (CA) in corporate social reporting (CSR): Revisiting the debate on the units of analysis and the ways to define them In *British Accounting Association Annual Conference 2007*, 3–5.

Further Reading

For more comprehensive coverage of manifest and latent content, including patterns and projections, see Potter and Levine-Donnerstein (1999). For more about content analysis using human coding, see Potter and Levine-Donnerstein (1999) and Franzosi (1989). For an in-depth book about content analysis, see Krippendorff (2012). For more about coding in qualitative research, see Saldaña (2015).

Chapter 3
Planning for Text Analytics

Abstract This chapter encourages readers to consider the reason for their analysis to chart the correct path for conducing it. This chapter outlines the process for planning the text analytics process. The chapter starts by asking the analyst to consider the objective, data availability, cost, and outcome desired. Analysis paths are then shown as possible ways to achieve the goal.

Keywords Text analytics · Text mining · Planning · Sampling

3.1 Introduction

The availability and accessibility of data have increased immensely with the Internet of Things (IoT) (Neuendorf 2016). As the amount of content increases, creating a new conceptual realm called "big data," traditional as well as more progressive content analysis approaches have been used in many fields of research and application. Not only has the amount of content increased, but also the nature of the content has changed. Through real-time communication and interactions via text, email, and social media, people can both generate and consume data in real time.

Technology-enabled analysis paved the way for an increase in text analysis applications across many industries and areas of research. Computer-aided text analysis (CATA) has allowed both analysts and non-analysts to combine qualitative and quantitative analysis methods and processes in modern text analytics applications. CATA is defined as "any technique involving the use of computer software for systematically and objectively identifying specified characteristics within text in order to draw inferences from text" (Kabanoff 1996, p. 507). The digital age intensified an already growing trend in the late 1990s toward more practical and applied approaches to content analysis (Stepchenkova 2012; Stone 1997). Content analysis refers to the systematic measurement of text or other symbolic materials (Krippendorff 2012; Shapiro and Markoff 1997). Content analysis and text analysis are often used interchangeably.

© Springer Nature Switzerland AG 2019
27
M. Anandarajan et al., *Practical Text Analytics*, Advances in Analytics and Data Science 2, https://doi.org/10.1007/978-3-319-95663-3_3

Text mining has become an all-encompassing term referring to all of the underlying processes of text analysis (Sebastiani 2002). In text mining, text analytics, and content analysis, text is the primary source of content and information. The similarities between and complementary nature of the two approaches to analyzing text have been detailed in recent literature. For instance, Yu et al. (2011) suggest that text mining is compatible with qualitative research, asserting that text mining and content analysis are similar. On the other hand, many researchers have shown how the two methods are complementary. According to Wiedemann (2013), computer-assisted text analytics can be used to support qualitative research. Lin et al. (2009) present an application using text mining as an enabler of content analysis for the classification of genres.

3.2   Initial Planning Considerations

The planning stage determines the rest of the analytics process, because it is at this point that the analyst makes important basic decisions. Regardless of the reason for the analysis, by the end of this stage, the analyst will have collected the data and defined the variables of interest. At this point, the analyst has at least a conception of the goal of the analysis in mind. In any text analytics application, there are four major considerations that will influence each step of the planning process, as outlined in Fig. 3.1.

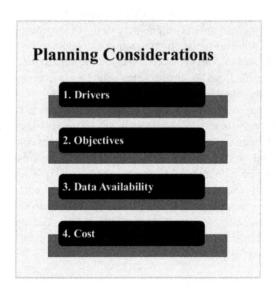

Fig. 3.1 Four initial planning considerations

3.2.1   Drivers

The process of a text analytics project will vary depending on the motivation behind the analysis. More specifically, the starting point of the analysis will influence the process and methods. Krippendorff and Bock (2009) suggest that there are three drivers of content analysis applications: (1) the text, (2) the problem, and (3) the method. We use this framework to describe the drivers of real-world text analytics applications. In text-driven analysis, the analysis is undertaken because of access to and interest in a dataset. The data may be information the analyst owns or collected, or data that were generated by and/or acquired in an alternate way. For instance, a company's marketing department may conduct a text-driven social media-based market research project to determine the sentiment of its customers regarding a new product or service, without having any preconception about what the true sentiment may be. The company may collect the data automatically and provide them to the analyst.

Unlike text-driven analysis, problem-driven analysis relies on background theory and may include the testing of a hypothesis using a representative sample. Problem-driven analysis for hypothesis testing is akin to content analysis for deductive inference. In text analytics, on the other hand, the basis for testing a hypothesis may not be as rigid. In many business applications, problem-driven analysis can be used to compare past performance or organizational beliefs about consumers. As an example, a company may have reason to believe that the sentiment about one of its products is positive by looking at the average number of stars assigned on online consumer reviews. A sentiment analysis could be undertaken to compare customer sentiment to star ratings.

Finally, method-driven analysis is application based and undertaken with techniques, software, or methodology in mind. As new methodologies and tools are applied to text, method-driven analysis may be used to test new methods or evaluate new ways of doing things. For example, an analyst may want to demonstrate the benefits of performing a cluster analysis using complaint narrative data to categorize customer complaints that are missing categorical information due to technical errors. While the focus is on the application, the analyst must still ensure that the method is appropriate and methodologically sound.

3.2.2   Objectives

The objective of the analysis will inform the type of data, methods and resources necessary to complete the analysis. In most business applications of text analytics, the overarching goal of the analysis is to derive value from the data resources. This value-added activity can help businesses make decisions and gain competitive advantages (Ur-Rahman and Harding 2012). Potential business text analytics project objectives include describing or classifying text data; identifying patterns, relationships, and networks; visualizing information; extracting or summarizing information; and applying state-of-the-art technologies or testing a hypothesis.

It is important to have a clear understanding of the objectives before beginning a text analytics project. As in any business project, well-defined objectives lead to measurable performance and benchmarking capabilities.

3.2.3 Data

Choosing the data to analyze is a key decision. If the goal of the analysis is analyzing data, that choice is quite simple. However, if the analysis is driven by methods or problems, it is important to identify the data that will meet the objective. Having made that decision, the next step is to determine whether the data are available. In some cases, such as survey research, new data may need to be gathered. The analyst may often be able to use data he or she already owns or can access. In other cases, the data may be available on the web or from other electronic sources.

In addition to data generation, which is covered in Sect. 3.5, understanding the data is crucial.

3.2.4 Cost

The final consideration is the cost of the analysis. If the analysis is data driven, the data already exist and will not add to the overall project cost. However, if data must be collected, the cost of doing so should be considered. Luckily, text data is usually an available but untapped organizational resource. In fact, the growing amount of data, both proprietary and public, has created an abundance rather than a shortage of material (Cukier 2010). Tapping into these available resources can significantly reduce the cost associated with the analysis.

If the analysis is method-driven, the technological resources needed to complete the analysis are already available. However, if the analysis is driven by data or a specific problem, technology costs should be considered when planning and managing a text analytics project. Commercial software programs can be costly. If cost is a concern, open-source or web-based software programs can provide an ideal cost-saving solution.

3.3 Planning Process

As Fig. 3.2 illustrates, the planning stage involves several major tasks: framing the problem, generating the data, and making decisions about the analysis.

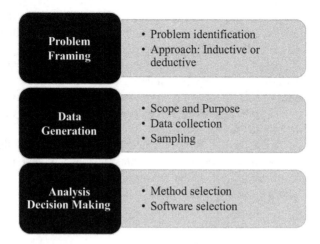

Fig. 3.2 Text analytics planning tasks

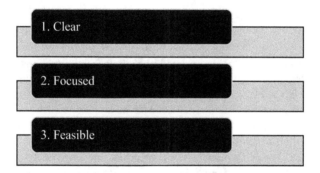

Fig. 3.3 Three characteristics of good research problems

3.4 Problem Framing

3.4.1 Identifying the Analysis Problem

Problem identification is an important part of the text analytics planning process. As in content analysis, text analytics is a research method. The research problem is the topic or focus of the analysis. As Fig. 3.3 notes, the research problem should be clear, focused, and feasible.

The text analytics focus should be clear and well defined. The research problem helps inform the population of text documents that will be researched. If the focus is not clear, the relevant population may be erroneously chosen, rendering the analysis results invalid. The analyst must make many important decisions with respect to the analysis, and a clear focus will make it easier to make these decisions.

To meet project deadlines and goals, it is important to narrow the focus of the analysis. Additionally, a project with a research question that is too broad may be difficult to manage and implement (Boudah 2011). For instance, if there is an organizational concern about customer satisfaction, measuring customer satisfaction from customer feedback might be proposed as the focus of the text analytics project. This is a broad focus, and the insights about the customers overall could miss important differences among customers. A clearer focus would be to investigate the sentiment of customers about a product or service in a given region. The results of an analysis with this research focus can inform marketing decisions and strategies for the product in that region.

Finally, the analysis should be feasible, meaning it should be possible to answer the research question. It is important to set the text analytics project up for success at the beginning of the project. If the research question cannot be answered, the focus of the analysis should be altered prior to the analysis.

3.4.2 Inductive or Deductive Inference

Similar to content analysis, the objective of the text analytics application will help to determine the role of background information, domain knowledge, and theory. If the goal is to learn something new from the text data, as in content analysis for inductive inference, it may be reasonable to keep external influences, such as background and theory, to a minimum. On the other hand, text analytics projects can help expand a business' current knowledge base, and the use of domain knowledge and existing organizational knowledge can be invaluable. Akin to content analysis for deductive inference, existing data, information, and theory can be used for hypothesis testing, benchmarking, and model building.

As in deductive content analysis, basic information about the analysis planned is vital. In problem-driven text analytics, it is important to collect background information about the problem, because it is the motivation for the project. In many cases, this step may include gathering existing theoretical information or even collecting supporting data. In problem-driven text analytics, domain knowledge can also contribute to the formulation of hypotheses, whether formal or informal.

Similar to inductive content analysis, the planning stage in text-driven analytics focuses on the data itself. The analyst may perform an initial exploratory data analysis to become better acquainted with the data. If quantitative data are also present, such an analysis may help the analyst better understand the quantitative variables and how they relate to the qualitative data.

3.5 Data Generation

3.5.1 Definition of the Project's Scope and Purpose

In text analytics, each record of the collected textual data should have a unique identifier that can be used to refer to that instance. These instances are referred to as documents. Documents will sometimes correspond to physical documents, as in the analysis of written materials such as press briefings or corporate reports (Feldman and Sanger 2007). In some cases, the documents will not necessarily take the form of a physical document, as is the case of individual tweets, consumer reviews, or messages. All of these documents comprise a document collection.

As in content analysis, a major decision when planning a text analytics project is choosing the scope of the analysis. We want to determine the unit of analysis and granularity of this unit. In this case, it may be documents, sentences, phrases, words, or characters. It is more common in natural language processing than in text analytics to analyze text data at the word or character level. In text analytics, the focus is typically at the document level. We want to be able to describe or categorize a document based on the words, themes, or topics present in the document.

In defining the scope of the analysis, we also want to identify the information in the documents in which we are most interested, whether it is document categories, meaning, or content. If we want to group or categorize documents, our methods will be different than if we want to learn about the meaning of the documents. With respect to the content of the documents, we may be interested in the semantic or sentiment information. The semantic information derived from documents tells us how the words in the documents relate to one another, such as co-occurrences (Griffiths et al. 2007). Sentiment information concerns feelings, emotions, and polarity based on the words in the documents.

3.5.2 Text Data Collection

As with any data analysis, analysis using text data is only as good as the data being used as input to the analysis. For this reason, we prefer to collect and use high-quality data. While data quality can be judged with regard to many aspects, in Fig. 3.4 we highlight some important dimensions of quality from Pipino et al. (2002) that are relevant for text data: consistency, relevancy, completeness, ease of manipulation, free of error, and interpretability.

We want our text data to be relevant to our objective and area of interest. If our data are not relevant to the research objective, our analysis results will not be valid. We can ensure that we are using relevant data by carefully selecting the documents in our document collection. In choosing these relevant documents, we want to be certain that we can interpret the text and understand any special formatting. For instance, if we are using text data from Twitter, "#" will have a different meaning than in text data from newspaper articles. Any special rules or necessary domain knowledge should be considered to create a high-quality document collection.

In order to analyze a dataset, we need to be able to manipulate it, not only to improve the quality but also to preprocess it. Ease of manipulation is especially important when we want to improve the quality of our data, especially with respect to insuring that it is free of error, consistent, and complete. We can improve the quality of our data by performing data cleansing or scrubbing to remove duplicates and errors (Rahm and Do 2000). This step may require removing inconsistencies and incomplete documents. Initial consideration of the quality of the text data can

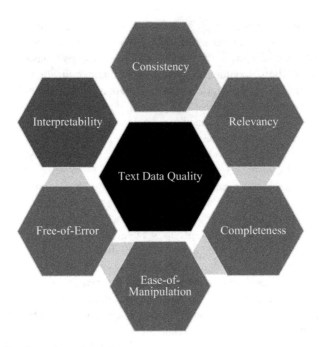

Fig. 3.4 Quality dimensions of text data

save us time in preparation and preprocessing. If the analysis is not text driven, the text data will need to be collected or generated. Text data can come from one or many sources and can be manually or automatically collected.

The unstructured nature of text data can make the text data collection process more difficult than with other types of data. Unlike other types of information, text data are created and consumed by humans (Zhai and Massung 2016). Research in the social sciences may include data from surveys, interviews, or focus groups (Granello and Wheaton 2004). In this area, the document collection will likely be smaller than other types of data. Problem-driven analyses, such as those in business environments, may utilize proprietary electronic data, such as that generated by an enterprise resource planning (ERP) or customer relationship management (CRM) system and stored in a database or data warehouse. In this case, the analyst will need to retrieve the relevant data for the text analytics project by using a querying language such as SQL.

Another source of relevant data may be the web. Web-based data are becoming increasingly popular in text analytics applications because of their many benefits, including availability, accessibility, and low cost. This data may already be available in a convenient format, such as via file transfer protocol (FTP) or on a webpage in .csv, Excel, or text format. In other cases, access to structured data may be available through an application programming interface (API), which requires some understanding of programming. If the data are contained within a webpage or many webpages, web scraping can be used to automatically collect the data.

This method works well with social media, discussion boards, blogs, and feeds (Webb and Wang 2014). In text analytics applications, especially those using web data, it is common to collect and use data from more than one source. In this case, care must be taken to standardize the data to make it as uniform as possible to maintain the quality of the data.

When collecting text data, we will often collect supporting data, known as metadata. This metadata will sometimes include a time stamp, unique identification number, and other classifying or identifying information for that document. For many analysis methods, metadata can and should be incorporated into the analysis. For instance, one analysis method covered in Chap. 9, classification analysis, requires a classification for each document. This information should be retained when collecting the data and stored as metadata.

3.5.3   Sampling

A population is a collection of something of interest, such as tweets or blog posts (Scheaffer et al. 2011). A population includes every instance of interest in the research. In planning the text analysis, the analyst should consider the population to be studied. In many cases, the entire population is much too large to collect and analyze. For this reason, we aim to create a sample of this population, which can be used to represent or generalize to the entire population. A sample is a smaller subset of the larger population. We consider two methods: non-probability and probability sampling.

3.5.3.1   Non-probability Sampling

In non-probability sampling methods, observations from the target population are chosen for the sample in a nonrandom fashion. Non-probability sampling methods are prominent in content analysis, due to its qualitative nature. We present two non-probability sampling methods: convenience and relevance.

Convenience sampling is the least restrictive sampling type, in which the availability of the sample determines inclusion (Marshall 1996). Convenience sampling is an inexpensive sa-mpling option; however, analysis using this method may suffer from a lack of generalizability and bias. While convenience sampling suffers from its subjectivity, in cases where a homogenous group is being studied, convenience sampling is appropriate (Webb and Wang 2014).

In relevance or purposive sampling, text data observations are chosen for inclusion in the sample if their inclusion contributes to answering the research question. Relevance sampling requires the analyst to choose the text documents to include, meaning that the analyst must understand and consider each of the documents (Krippendorff 2004). For smaller samples, this approach may be feasible; however, it can be costly and time consuming for wider-scale analyses.

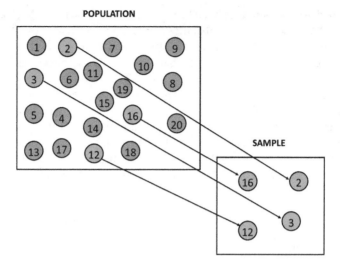

Fig. 3.5 Simple random sampling

3.5.3.2   Probability Sampling

In a simple random sample, a set number of observations are chosen for inclusion in the sample at random, using either a random number generator or an alternative computer-aided method. In a simple random sample in text analytics, each document has the same, or equal, probability of inclusion in the sample. In Fig. 3.5, a simple random sample of 4 documents is created from the population of 20 documents. The four documents were chosen using a random number generator.[1]

In systematic sampling, every nth observation is chosen for inclusion in the sample (Krippendorff 2012). In the systematic sampling example in Fig. 3.6, every fifth document is selected for inclusion in the sample of four documents.

A stratified random sample, shown in Fig. 3.7, acknowledges that subgroups exist in the data and those subgroups are separately sampled and later combined to form a single sample (Krippendorff 2012).

3.5.3.3   Sampling for Classification Analysis

In classification analysis, introduced in Chap. 9, the analysis sample is split into two smaller samples, the training and testing set, prior to the analysis. The training set is used to build the classification model, and the testing set is used to determine the accuracy of the model in making predictions. For this reason, sampling for this type of analysis deserves additional consideration. In sampling for classification

---

[1] In Microsoft Excel, random numbers can be generated using the function = RANDBETWEEN. The function requires minimum and maximum values as inputs. In the example the function would be = RANDBETWEEN(1,20), and the function would need to be copied to four cells to produce four random numbers between 1 and 20.

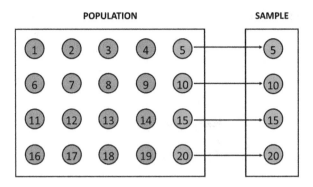

Fig. 3.6 Systematic sampling

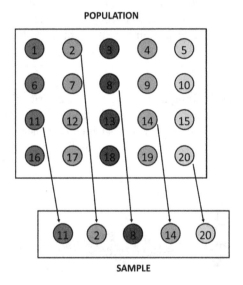

Fig. 3.7 Stratified sampling

analysis, there are two sampling methods proposed by Yang (1996): proportion-enforced and completeness-driven sampling. In proportion-enforced sampling, a systematic sample reflecting the training data is used as the training sample. On the other hand, in completeness-driven sampling, there is more control over the category observations included in the training set.

3.5.3.4 Sample Size

The sample should be large enough to be representative of the larger dataset. In quantitative analysis, there are guidelines for choosing the sample size. However, determinations about the sample size for a qualitative analysis can be based on published sample sizes or the sample sizes used in previous analyses (Webb and Wang 2014). In general, the larger the sample, the smaller the sampling error; however, the relationship between size and error represents diminishing returns (Marshall 1996). Put simply, a larger sample is preferable to a smaller sample, but an unnecessarily large sample can increase costs and inefficiency without realizing notable gains in error reduction.

3.6   Method and Implementation Selection

3.6.1   Analysis Method Selection

Unless the research is method-driven, the most relevant analysis method must be chosen based on the objective and scope of the project. Figure 3.8 outlines some simple questions that can help determine the type of analysis that should be used. By now, the analyst should have a fairly well-established idea of the objective and scope of the analysis, including its focus. Determining if the focus is on the categorization of documents or the meaning of documents can help narrow down the analysis method options.

If the focus of the analysis is to categorize or classify the documents in the document collection, the analysis method will depend on whether or not the available data have category or class labels. If such labels exist, classification analysis can be used, which has the ability to make predictions, as described in Chap. 9. If

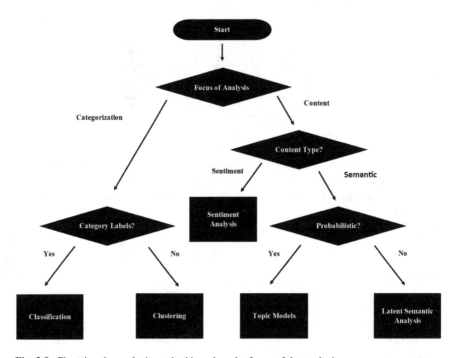

Fig. 3.8 Choosing the analysis method based on the focus of the analysis

labeled data are not available, cluster analysis, covered in Chap. 7, can be used to group documents into naturally occurring clusters.

On the other hand, the analysis may be concerned with the content of the documents. To identify the sentiment of the documents in the document collection, sentiment analysis can be performed, which is presented in Chap. 10. As demonstrated in this chapter, if the data have sentiment labels, classification methods can also be used to classify the documents by sentiment. To identify semantic information in the documents, there are two analysis methods available: latent semantic analysis (LSA) and topic models. Topic models are probabilistic, generative models, while LSA represents documents and words in vector space through dimension reduction methods. LSA is covered in Chap. 6, and topic models are discussed in Chap. 8.

3.6.2 The Selection of Implementation Software

Once the analysis method is chosen, it is important to consider how the analysis will be performed and which software will be used. In choosing software, some important considerations are cost, functionality, and capability. For instance, if cost is a concern, an open-source program, rather than commercial software, is preferable. With respect to functionality, the chosen software should be a good match with the analyst's skill set or should be user-friendly enough in the case of a learning curve. Finally, the analyst should ensure that the software can conduct the planned analysis. In Chaps. 13, 14, 15 and 16, we present real-world use cases with four popular software programs, including both commercial and open-source programs, for text data analysis.

There are many packages available in the R open-source software for the analysis of text, including tm (Feinerer et al. 2008) and tidytext (Silge and Robinson 2016). Chapter 13 presents a sentiment analysis application using the tidytext package. Python is an open-source software program with packages such as NLTK, text mining, pattern, and NumPy that can be used for text preprocessing, natural language processing, and text analytics (Bird et al. 2009). Chapter 14 describes an application of latent semantic analysis using Python. Rapidminer Text Mining and sentiment analysis can be used for text extraction and analysis, with an emphasis on mining web data. Chapter 15 provides an example of classification analysis using Rapidminer. SAS Text Miner (Zanasi 2005) is a commercial software program that uses SAS Enterprise Miner for the analysis of text data. Chap. 16 presents a visualization application using SAS Visual Text Analytics.

Key Takeaways
- Four important initial considerations when planning a text analytics project are drivers, objectives, data, and cost.
- The planning process can be broken down into three major planning tasks: problem framing, data generation, and analysis decision-making.

References

Bird, S., Klein, E., & Loper, E. (2009). *Natural language processing with Python: analyzing text with the natural language toolkit.* O'Reilly Media, Inc.

Boudah, D. J. (2011). Identifying a research problem and question and searching relevant literature. In *Conducting educational research: Guide to completing a major project.* Thousand Oaks: SAGE Publications.

Cukier, K. (2010). Data, data everywhere: A special report on managing information. *Economist Newspaper.*

Feinerer, I., Hornik, K., & Meyer, D. (2008). Text Mining Infrastructure in R. *Journal of Statistical Software, 25*(5): 1–54. http://www.jstatsoft.org/v25/i05/.

Feldman, R., & Sanger, J. (2007). *The text mining handbook: Advanced approaches in analyzing unstructured data.* Cambridge: Cambridge University Press.

Granello, D. H., & Wheaton, J. E. (2004). Online data collection: Strategies for research. *Journal of Counseling & Development, 82*(4), 387–393.

Griffiths, T. L., Steyvers, M., & Tenenbaum, J. B. (2007). Topics in semantic representation. *Psychological Review, 114*(2), 211–244.

Kabanoff, B. (1996). Computers can read as well as count: How computer-aided text analysis can benefit organisational research. *Trends in organizational behavior, 3*, 1–22.

Krippendorff, K. (2004). Reliability in content analysis: Some common misconceptions and recommendations. *Human communication research, 30*(3), 411–433.

Griffiths, T. L., Steyvers, M., & Tenenbaum, J. B. (2007). Topics in semantic representation. *Psychological Review, 114*(2), 211–244.

Krippendorff, K. (2012). *Content analysis: An introduction to its methodology.* Thousand Oaks: Sage.

Krippendorff, K., & Bock, M. A. (2009). *The content analysis reader.* Thousand Oaks: Sage.

Kroenke, D. M., & Auer, D. J. (2010). *Database processing* (Vol. 6). Upper Saddle River: Prentice Hall.

Lin, F. R., Hsieh, L. S., & Chuang, F. T. (2009). Discovering genres of online discussion threads via text mining. *Computers & Education, 52*(2), 481–495.

Marshall, M. N. (1996). Sampling for qualitative research. *Family Practice, 13*(6), 522–526.

Neuendorf, K. A. (2016). *The content analysis guidebook.* Sage.

Pipino, L. L., Lee, Y. W., & Wang, R. Y. (2002). Data quality assessment. *Communications of the ACM, 45*(4), 211–218.

Rahm, E., & Do, H. H. (2000). Data cleaning: Problems and current approaches. *IEEE Data Engineering Bulletin, 23*(4), 3–13.

Scheaffer, R. L., Mendenhall, W., III, Ott, R. L., & Gerow, K. G. (2011). *Elementary survey sampling.* Boston: Cengage Learning.

Scheaffer, R. L., Mendenhall, W., III, Ott, R. L., & Gerow, K. G. (2011). *Elementary survey sampling.* Boston: Cengage Learning.

Sebastiani, F. (2002). Machine learning in automated text categorization. *ACM computing surveys (CSUR), 34*(1), 1–47.

Shapiro, G., & Markoff, J. (1997). A Matter of Definition. In C.W. Roberts (Ed.), *Text Analysis for the Social Sciences: Methods for Drawing Statistical Inferences from Texts and Transcripts,* Mahwah, NJ: Lawrence Erlbaum Associates.

Silge, J., & Robinson, D. (2016). tidytext: Text Mining and Analysis Using Tidy Data Principles in R. *Journal of Statistical Software, 1*(3).

Stepchenkova, S. (2012). Content analysis. In L. Dwyer et al. (ed.), *Handbook of research methods in tourism: Quantitative and qualitative approaches* (pp. 443–458). Edward Elger Publishing.

Stone, P.J. (1997). Thematic text analysis. In C.W. Roberts (Ed.), *Text Analysis for the Social Sciences: Methods for Drawing Statistical Inferences from Texts and Transcripts* (pp. 35-54). Mahwah, NJ: Lawrence Erlbaum Associates.

Ur-Rahman, N., & Harding, J. A. (2012). Textual data mining for industrial knowledge management and text classification: A business oriented approach. *Expert Systems with Applications, 39*(5), 4729-4739.

Webb, L. M., & Wang, Y. (2014). Techniques for sampling online text-based data sets. In *Big data management, technologies, and applications* (pp. 95–114). Hershey: IGI Global.

Wiedemann, G. (2013). Opening up to big data: Computer-assisted analysis of textual data in social sciences. *Historical Social Research/Historische Sozialforschung, 38*(4), 332–357.

Yang, Y. (1996). Sampling strategies and learning efficiency in text categorization. In M. Hearst & H. Hirsh (Eds.), *AAAI spring symposium on machine learning in information access* (pp. 88–95). Menlo Park: AAAI Press.

Yu, C. H., Jannasch-Pennell, A., & DiGangi, S. (2011). Compatibility between text mining and qualitative research in the perspectives of grounded theory, content analysis, and reliability. *The Qualitative Report, 16*(3), 730.

Zanasi, A. (2005). Text mining tools. In Text Mining and its Applications to Intelligence, *CRM and Knowledge Management*. WIT Press, Southampton Boston, 315–327.

Zhai, C., & Massung, S. (2016). *Text data management and analysis: A practical introduction to information retrieval and text mining*. San Rafael: Morgan & Claypool.

Further Reading

For more thorough coverage of the research problem and question, see Boudah (2011). Database management, processing, and querying are beyond the scope of this book. For more comprehensive coverage of these topics, see Kroenke and Auer (2010). Web scraping is very important, but also beyond the scope of this book. For more detailed information and instructions, see Munzert et al. (2014) for web scraping using R or Mitchell (2015) for web scraping using Python.

Part II
Text Preparation

Chapter 4
Text Preprocessing

Abstract This chapter starts the process of preparing text data for analysis. This chapter introduces the choices that can be made to cleanse text data, including tokenizing, standardizing and cleaning, removing stop words, and stemming. The chapter also covers advanced topics in text preprocessing, such as n-grams, part-of-speech tagging, and custom dictionaries. The text preprocessing decisions influence the text document representation created for analysis.

Keywords Text preprocessing · Text parsing · n-grams · POS tagging · Stemming · Lemmatization · Natural language processing · Tokens · Stop words

4.1 Introduction

By the end of the planning stage, the data should be collected, and the goal of the analysis should be well defined. After completing the planning stage, the next step is to prepare the data for analysis. Each record of the collected text data should have a unique identifier that can be used to refer to that instance. In text analytics, these instances are known as documents. A document is typically made up of many characters. The many documents make up a document collection or corpus. Characters are combined to form words or terms in a given language. These words are the focus of our analysis, although groupings of terms can also be the chosen unit of analysis, as described in this chapter. The collection of terms is sometimes called the vocabulary or dictionary. Figure 4.1 illustrates the components of our text data.

Let's consider an example of a document collection in which ten people were told to envision and describe their dog. The dog could be wearing dog clothes. Some of these people describe their own dogs, while others describe a fictional dog. The document collection is shown in Fig. 4.2.

© Springer Nature Switzerland AG 2019

M. Anandarajan et al., *Practical Text Analytics*, Advances in Analytics and Data Science 2, https://doi.org/10.1007/978-3-319-95663-3_4

Fig. 4.1 Hierarchy of
terms and documents

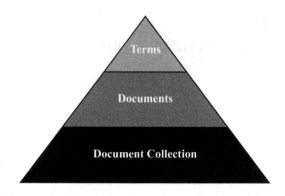

4.2   The Preprocessing Process

Much of the text preparation and preprocessing methods have their roots in natural
language processing. Text preprocessing takes an input of raw text and returns
cleansed tokens. Tokens are single words or groups of words that are tallied by their
frequency and serve as the features of the analysis.

The preprocessing process includes (1) unitization and tokenization, (2) standardization and cleansing or text data cleansing, (3) stop word removal, and (4)
stemming or lemmatization. The stages along the pipeline standardize the data,
thereby reducing the number of dimensions in the text dataset. There is a balance
between retained information and reduced complexity in the choices made during
the process. This process is depicted in Fig. 4.3.

Each step removes unnecessary information from the original text. Proper preprocessing of text data sets up the analysis for success. In text analysis, far more
time is spent in preparing and preprocessing the text data than in the analysis itself
(Dumais et al. 1998). Diligent and detailed work in cleansing and preprocessing
makes the analysis process smoother.

4.3   Unitize and Tokenize

The first step involves the choice of the unit of text to analyze and the separation of
the text based on the unit of analysis. This unit could be a word; however, in other
cases, it may be a grouping of words or a phrase. Single words are the simplest
choice and make a good starting point. It is difficult for a computer to know where
to split the text. Fortunately, most text mining software contains functions to split
text, because computers do not naturally sense when punctuation designates the end
of a word or sentence. For example, apostrophes could indicate the end of a token,
or not, depending on the use (Weiss et al. 2010).

In our example, tokenization is done under the assumption of bag-of-words
(BOW), meaning that the grammar and ordering of the text in a document is not

DOCUMENT 1
My favorite dog is fluffy and tan.

DOCUMENT 2
the dog is brown and cat is brown

DOCUMENT 3
My favorite hat is brown and coat is pink

DOCUMENT 4
My dog has a hat and leash. ♥

DOCUMENT 5
He has a fluffy coat and brown coats.

DOCUMENT 6
The dog is brown and fluffy & has a brown coat.

DOCUMENT 7
MY dog is white with brown spots.

DOCUMENT 8
The white dog has a Pink coat and the Brown dog is fluffy

DOCUMENT 9
The 3 fluffy dogs AND 2 brown hats are my favorites!

DOCUMENT 10
MY fluffy dog has a white coat and hat .

Fig. 4.2 Example document collection

considered in building the quantitative representation of the qualitative text data. First, we analyze the text in the document collection at the word or term level. For instance:

Document 1: *My Favorite Dog Is Fluffy and Tan* There are seven words in this document. Next, through tokenization, we separate the text into a more usable form, known as tokens.

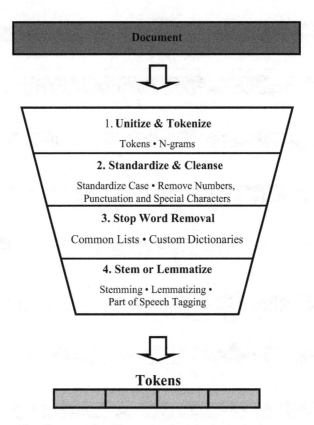

Fig. 4.3 The text data pre-processing process

Based on Document 1, we have eight tokens. Each of the seven words is a token, in addition to the period at the end of the sentence. The tokenized documents are shown in Fig. 4.4. Each bracket in the documents represents a token.

4.3.1   N-Grams

N-grams are an alternative to single words in the tokenization process. N-grams are tokens; they are consecutive word sequences with length n. For instance, bigrams are tokens composed of two side-by-side words; a single word is known as a unigram. N-grams retain information about the co-occurrence of words, because they group adjacent words into the same token.

Visualize the process as a picture frame that contains n words. Initially, the frame rests over the first n words. This counts as a token. The frame then moves over one word resulting in the exclusion of the first word. This is the second token. This process repeats for the length of the text (Struhl 2015). We will demonstrate this process with n equal to two, known as a bigram. Again, we use Document 1.

DOCUMENT 1
[My] [favorite] [dog] [is] [fluffy] [and] [tan] [.]

DOCUMENT 2
[the] [dog] [is] [brown] [and] [cat] [is] [brown]

DOCUMENT 3
[My] [favorite] [hat] [is] [brown] [and] [coat] [is] [pink] [.]

DOCUMENT 4
[My] [dog] [has] [a] [hat] [and] [leash] [.] [♥]

DOCUMENT 5
[He] [has] [a] [fluffy] [coat] [and] [brown] [coats] [.]

DOCUMENT 6
[The] [dog] [is] [brown] [and] [fluffy] [&] [has] [a] [brown] [coat] [.]

DOCUMENT 7
[MY] [dog] [is] [white] [with] [brown] [spots] [.]

DOCUMENT 8
[The] [white] [dog] [has] [a] [Pink] [coat] [and] [the] [Brown] [dog] [is] [fluffy]

DOCUMENT 9
[The] [3] [fluffy] [dogs] [AND] [2] [brown] [hats] [are] [my] [favorites] [!]

DOCUMENT 10
[MY] [fluffy] [dog] [has] [a] [white] [coat] [and] [hat] [.]

Fig. 4.4 Tokenized example documents

The first token is:
| *My favorite* | *dog is fluffy and tan.*
The second token is:
My | *favorite dog* | *is fluffy and tan.*
The final token is:
My favorite dog is fluffy and | *tan.* |
The bigram representation of Document 1 is:

My favorite
favorite dog
dog is
is fluffy
fluffy and
and tan
tan

This procedure can be used to create *n*-grams for higher values of *n*.

4.4   Standardization and Cleaning

First, we want to standardize and clean the tokens. These transformations level the playing field by making the terms in each of the documents comparable. For instance, we do not want *character*, *character*, and *Character* to be considered separate items just because one has a comma, and another has an upper case *C*. This standardization and cleaning prevents this possibility from occurring.

Our first step is to convert the terms in the text to lower case. In this step, any capital letters are converted to lower case. Without this conversion, the first token in Document 1, *My*, and the first token in Document 7, *MY*, would erroneously be considered two different terms.

Following this conversion, we want to remove numbers, punctuation, and special characters. In Document 9, we will remove the numbers 3 and 2. We will also remove any punctuation at the ends of sentences, such as periods and exclamation points. We remove periods from Documents 1, 3, 4, 5, 6, 7, and 10. We also remove an exclamation point from Document 9 and an ampersand from Document 6. In this document collection, we have one special character, a ♥ in Document 4, which is removed. In real-world text data, there may be additional characters and tokens to clean. For instance, in some text documents, there may be extra spaces, such as white space. These are also eliminated at this stage. The results of cleansing and standardizing our text data appear in Fig. 4.5.

4.5   Stop Word Removal

Next, we want to drop frequently used filler words, or stop words, which add no value to the analysis. According to the Oxford English Dictionary, *and*, *the*, *be*, *to*, and *of* are the most common words in the English language.[1] In the case of text analysis, we remove common terms because, although common terms such as

---

[1] https://en.oxforddictionaries.com/explore/what-can-corpus-tell-us-about-language

DOCUMENT 1
[my] [favorite] [dog] [is] [fluffy] [and] [tan]

DOCUMENT 2
[the] [dog] [is] [brown] [and] [cat] [is] [brown]

DOCUMENT 3
[my] [favorite] [hat] [is] [brown] [and] [coat] [is] [pink]

DOCUMENT 4
[my] [dog] [has] [a] [hat] [and] [leash]

DOCUMENT 5
[he] [has] [a] [fluffy] [coat] [and] [brown] [coats]

DOCUMENT 6
[the] [dog] [is] [brown] [and] [fluffy] [&] [has] [a] [brown] [coat]

DOCUMENT 7
[my] [dog] [is] [white] [with] [brown] [spots]

DOCUMENT 8
[the] [white] [dog] [has] [a] [pink] [coat] [and] [the] [brown] [dog] [is] [fluffy]

DOCUMENT 9
[the] [fluffy] [dogs] [and] [brown] [hats] [are] [my] [favorites]

DOCUMENT 10
[my] [fluffy] [dog] [has] [a] [white] [coat] [and] [hat]

Fig. 4.5 Cleansed and standardized document collection

these serve a grammatical purpose, they provide little information in terms of content (Salton 1989; Wilbur and Sirotkin 1992).

A collection of stop words is known as a stop list. Alternatively, this collection of words can be known as a dictionary. There are several stop lists such as those presented in Chaps. 13, 14, 15 and 16 that are utilized by software programs for text analytics. These lists include many different languages and methods.[2] The

---

[2] Stop lists in more than 40 languages can be found at http://www.ranks.nl/stopwords

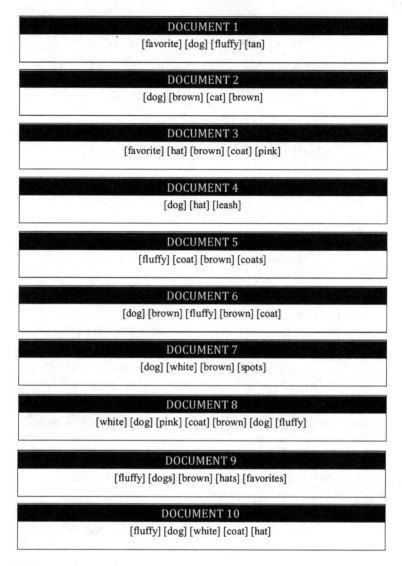

Fig. 4.6 Documents after stop word removal

SMART information retrieval system, introduced by Salton (1971), has a popular stop word list containing 571 words.[3]

In our example, we will use the SMART dictionary (Salton 1971) to identify and remove stop words from our documents. The following stop words were removed from the documents: *a, and, has, he, is, my, the,* and *was.* The resulting document collection after stop word removal is displayed in Fig. 4.6. An alternative to using existing stop word dictionaries is to create a custom dictionary.

---

[3] The stop word list based on the SMART information retrieval system can be found at http://www. ai.mit.edu/projects/jmlr/papers/volume5/lewis04a/a11-smart-stop-list/english.stop

4.5.1 Custom Stop Word Dictionaries

Standard stop word dictionaries eliminate many of the most common words in a given language, but sometimes we also want to drop domain- or project-specific words or tokens. In this case we can build a custom dictionary to complete this task. Often, projects will have topic-specific words that occur frequently but add little value. In the documents in our example, we may consider adding the term *dog* to a custom stop list. Given that the ten respondents were told to describe their dog, this word may not be informative in the analysis. Since one respondent described both a cat and a dog, in this case, we chose not to remove the term. If this were not the case, the word *dog* would be included in a custom stop word dictionary and removed.

To identify these words, we can also look at term frequencies. Words that have high frequencies across all documents in our collection but do not provide informational content are good candidates for removal. After making these choices, we can select a subset of the data with the term in it and read it. While reading, we want to ensure that the term does not provide information that is relevant to the analysis question. In the creation of a custom stop word dictionary, the process should be repeated, creating subsets for reading several times to check for multiple uses of the term (Inmon 2017).

Custom dictionaries can be used in many text analysis projects and not exclusively to filter stop words. Custom dictionaries are created in information retrieval to find keywords in context (KWIC). Instead of finding tokens for removal, these keywords are used as search terms. Additional uses of common and custom lexicons will be presented in Chap. 10 for sentiment analysis.

4.6 Stemming and Lemmatization

4.6.1 Syntax and Semantics

Prior to introducing the final preprocessing step, it is necessary to consider two important concepts: syntax and semantics. Syntax concerns sentence structure, including grammar and parts of speech. Parts of speech are grammatical categories or word classes, such as noun, verb, and adjective (Manning and Schütze 1999). Semantics, on the other hand, refers to meaning. Part-of-speech tagging is beneficial in text analysis because in identifying the part of speech of a token, the most likely meaning can also be identified.

Two important semantic concepts related to part-of-speech tagging are synonymy and polysemy. Synonymy refers to two different words having the same meaning. If a document were added that said, "My cap is brown," the terms *cap* and *hat* would demonstrate synonymy. Since the synonyms cannot be recognized automatically, each word is a separate token.

Polysemy refers to a single word having multiple meanings. In the case of the word *coat* in our example, in one context, it means a garment that can be worn, but in another context, it can refer to a dog's hair or fur. In our example, it is possible that in Document 10, which reads, "My fluffy dog has a white coat and hat," the dog is wearing a white coat, but it is also possible that the dog has white fur. As a noun, the word *coat* can mean "fur covering the body of an animal" (Wordnet 3.1).

4.6.2   Stemming

The final stage is to perform either stemming or lemmatization on the documents. Stemming and lemmatization involve breaking words down to their root word. Stemming involves the removal of a word's suffix to reduce the size of the vocabulary (Porter 1980). Lemmatization is similar to stemming, except it incorporates information about the term's part of speech (Yatsko 2011). Both methods combine words that contain the same root into a single token to reduce the number of unique tokens within the analysis set. Words with a common root often share a similar meaning. These words are then grouped into one token (Manning et al. 2008). There are exceptions to the roots of words sharing the same meaning, but the added reduction in complexity is often worth the price of incorrectly categorizing a few words.

As an example, let's use the root *train*. *Train* has several forms, including:

- *Train*
- *Trains*
- *Trained*
- *Training*
- *Trainer*

In stemming, these words return the root *train*. Common stemmers, such as Porter's (1980), Lovins' (1968), and Paice's (1990, 1994), use a series of rules to remove word endings. These algorithms aim to return the base word. The number of characters removed changes depending on the stemming algorithm. A stemmer that removes more from a word will result in less variation among tokens and more word forms grouped within the same token. Depending on the project, this could mean better results or increased errors (Manning et al. 2008).

As an example of stemming in our document collection, we can take a closer look at Document 9, shown in Fig. 4.7, before and after stemming.

The term *fluffy* describes the dog's fur or fluff. Using Porter's stemming algorithm (Porter 1980), the term *fluffy* is broken down to the root word *fluffi*. Other terms with this root will also be replaced by the root. Some terms that would also be truncated to the root *fluffi* are *fluffier*, *fluffiest*, *fluffiness*, and *fluffily*.

As the figure shows, in the document after stemming, the terms *dogs* and *hats* are converted to their singular form, *dog* and *hat*, respectively. The term *favorites* is not only broken down to its singular form but also further reduced to the root, *favorit*. Some terms that would also be stemmed to the root *favorit* are *favorite*, *favorites*, and *favorited*. The full, stemmed document collection appears in Fig. 4.8.

DOCUMENT 9 (BEFORE STEMMING)
[fluffy] [dogs] [brown] [hats] [favorites]

DOCUMENT 9 (AFTER STEMMING)
[fluffi] [dog] [brown] [hat] [favorit]

Fig. 4.7 Document 9 tokenized text before and after stemming

DOCUMENT 1
[favorit] [dog] [fluffi] [tan]

DOCUMENT 2
[dog] [brown] [cat] [brown]

DOCUMENT 3
[favorit] [hat] [brown] [coat] [pink]

DOCUMENT 4
[dog] [hat] [leash]

DOCUMENT 5
[fluffi] [coat] [brown] [coat]

DOCUMENT 6
[dog] [brown] [fluffi] [brown] [coat]

DOCUMENT 7
[dog] [white] [brown] [spot]

DOCUMENT 8
[white] [dog] [pink] [coat] [brown] [dog] [fluffi]

DOCUMENT 9
[fluffi] [dog] [brown] [hat] [favorit]

DOCUMENT 10
[fluffi] [dog] [white] [coat] [hat]

Fig. 4.8 Stemmed example document collection

4.6.3   Lemmatization

One difficulty encountered with stemming (and text analytics in general) is that a single word could have multiple meanings depending on the word's context or part of speech. Lemmatization deals with this problem by including the part of speech in the rules grouping word roots. This inclusion allows for separate rules for words with multiple meanings depending on the part of speech (Manning et al. 2008). This method helps improve the algorithm by correctly grouping tokens at the cost of added complexity.

As an example of lemmatization in our document collection, we can again look at Document 9 in Fig. 4.9. The figure depicts the document at the end of Step 3 in green and the document after stemming and after lemmatization in orange. As shown, stemming and lemmatization produce the same tokens for the terms *dog*, *brown*, and *hat* but vary with respect to *fluffy* and *favorite*.

Returning to the terms that would be truncated to the root *fluffi* using stemming, we can consider how lemmatization would impact them. These terms and their parts of speech are displayed in Table 4.1. As shown, all adjectives are lemmatized to *fluffy*, while the noun and adverb, *fluffiness* and *fluffily*, remain unchanged.

The same procedure can be done for the related terms that reduce to the root *favorit* in stemming. Table 4.2 displays the words, parts of speech, and

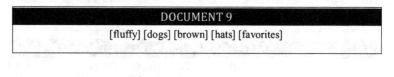

| DOCUMENT 9 |
| [fluffy] [dogs] [brown] [hats] [favorites] |

| DOCUMENT 9 (AFTER STEMMING) |
| [fluffi] [dog] [brown] [hat] [favorit] |

| DOCUMENT 9 (AFTER LEMMATIZATION) |
| [fluffy] [dog] [brown] [hat] [favorite] |

Fig. 4.9   Document 9 before and after stemming and lemmatization

Table 4.1   Document 9 related words, POS, and lemmatization for the word fluffy

Word	Part of speech	Lemmatization
Fluffy	Adjective	Fluffy
Fluffier	Adjective	Fluffy
Fluffiest	Adjective	Fluffy
Fluffiness	Noun	Fluffiness
Fluffily	Adverb	Fluffily

Table 4.2 Document 9 related words, POS, and lemmatization for the word favorite

Word	Part of speech	Lemmatization
Favorite	Noun/adjective	Favorite
Favorites	Noun	Favorite
Favorited	Verb	Favorited

DOCUMENT 1

[favorite] [dog] [fluffy] [tan]

DOCUMENT 2

[dog] [brown] [cat] [brown]

DOCUMENT 3

[favorite] [hat] [brown] [coat] [pink]

DOCUMENT 4

[dog] [hat] [leash]

DOCUMENT 5

[fluffy] [coat] [brown] [coat]

DOCUMENT 6

[dog] [brown] [fluffy] [brown] [coat]

DOCUMENT 7

[dog] [white] [brown] [spot]

DOCUMENT 8

[white] [dog] [pink] [coat] [brown] [dog] [fluffy]

DOCUMENT 9

[fluffy] [dog] [brown] [hat] [favorite]

DOCUMENT 10

[fluffy] [dog] [white] [coat] [hat]

Fig. 4.10 Lemmatized example document collection

lemmatization of these terms. As shown, *favorite* and *favorites*, which are primarily nouns, are lemmatized to *favorite*, while the verb, *favorited*, remains unchanged.

The full, lemmatized document collection is displayed in Fig. 4.10.

The choice between stemming and lemmatization is up to the analyst and will depend on the application and text data.

4.6.4   Part-of-Speech (POS) Tagging

Part-of-speech tagging involves labeling tokens or words by their part of speech (Manning and Schütze 1999). Two of the most popular tag sets in English are the Brown Corpus (Kučera and Francis 1967) and the Lancaster-Oslo-Bergen (LOB) Corpus (Johansson et al. 1978). A newer tag set, the Penn Treebank, was developed in 1989 and has over 7 million words tagged by their parts of speech (Taylor et al. 2003).

Part-of-speech tagging can be completed using one of the software programs described in Chap. 1, including many of those presented in Chaps. 13, 14, 15 and 16. In these programs, the documents are entered as inputs. The program processes and outputs the word and annotates the parts of speech (Bird et al. 2009). The method used to identify the part of speech may be rule-based, Markov model-based, or maximum entropy-based (Indurkhya and Damerau 2010). Additionally, machine learning techniques, such as those introduced in Chap. 9, can be used to automatically identify the parts of speech of words in the document collection. These methods can prevent errors caused by stemming or lemmatizing words to the same root that actually have different meanings depending on the part of speech, demonstrating polysemy. The accuracy of the analysis can be improved by grouping more words, such as synonyms, appropriately.

Key Takeaways
- The text preprocessing process involves unitization and tokenization, standardization and cleaning, stop word removal, and lemmatization or stemming.
- A custom stop word dictionary can be created to eliminate noise in the text.
- Part-of-speech tagging involves labeling tokens or words by their part of speech and can be used to prevent stemming and lemmatization-related errors.
- N-grams are consecutive token sequences with length n that preserve token co-occurrence.

References

Bird, S., Klein, E., & Loper, E. (2009). *Natural language processing with Python: Analyzing text with the natural language toolkit.* Beijing: O'Reilly Media, Inc.

Dumais, S., Platt, J., Heckerman, D., & Sahami, M. (1998, November). Inductive learning algorithms and representations for text categorization. In *Proceedings of the seventh international conference on Information and knowledge management* (pp. 148–155). ACM.

Indurkhya, N., & Damerau, F. J. (Eds.). (2010). *Handbook of natural language processing* (Vol. 2). Boca Raton: CRC Press.

Inmon, B. (2017). *Turning text into gold: Taxonomies and textual analytics.* Bradley Beach: Technics Publications.

Johansson, S., Leech, G. N., & Goodluck, H. (1978). *The Lancaster-Oslo/Bergen Corpus of British English.* Oslo: Department of English: Oslo University Press.

Kučera, H., & Francis, W. N. N. (1967). *Computational analysis of present-day American English.* Providence: Brown University Press.

Lovins, J. B. (1968). Development of a stemming algorithm. *Mechanical Translation and Computational Linguistics, 11*(1–2), 22–31.

Manning, C. D., & Schütze, H. (1999). *Foundations of statistical natural language processing.* Cambridge: MIT Press.

Manning, C., Raghavan, P., & Schütze, H. (2008). *Introduction to information retrieval.* Cambridge: Cambridge University Press. https://doi.org/10.1017/CBO9780511809071.

Paice, C. D. (1990). Another stemmer. *ACM SIGIR Forum, 24*(3), 56–61.

Paice, C. D. (1994, August). An evaluation method for stemming algorithms. In *Proceedings of the 17th Annual International ACM SIGIR Conference on Research and Development in Information Retrieval* (pp. 42–50). Springer-Verlag New York, Inc.

Porter, M. F. (1980). An algorithm for suffix stripping. *Program, 14*(3), 130–137.

Salton, G. (1971). *The SMART retrieval system: Experiments in automatic document processing.* Englewood Cliffs: Prentice-Hall.

Salton, G. (1989). *Automatic text processing: The transformation, analysis, and retrieval of.* Reading: Addison-Wesley.

Struhl, S. (2015). *Practical text analytics: Interpreting text and unstructured data for business intelligence.* London: Kogan Page Publishers.

Taylor, A., Marcus, M., & Santorini, B. (2003). The penn treebank: An overview. In *Treebanks* (pp. 5–22). Dordrecht: Springer.

Weiss, S. M., Indurkhya, N., Zhang, T., & Damerau, F. (2010). *Text mining: predictive methods for analyzing unstructured information.* Springer Science & Business Media.

Wilbur, W. J., & Sirotkin, K. (1992). The automatic identification of stop words. *Journal of Information Science, 18*(1), 45–55.

Yatsko, V. A. (2011). Methods and algorithms for automatic text analysis. *Automatic Documentation and Mathematical Linguistics, 45*(5), 224–231.

Further Reading

For a more comprehensive treatment of natural language processing, see Indurkhya and Damerau (2010), Jurafsky and Martin (2014), or Manning and Schütze (1999).

Chapter 5
Term-Document Representation

Abstract This chapter details the process of converting documents into an analysis-ready term-document representation. Preprocessed text documents are first transformed into an inverted index for demonstrative purposes. Then, the inverted index is manipulated into a term-document or document-term matrix. The chapter concludes with descriptions of different weighting schemas for analysis-ready term-document representation.

Keywords Inverted index · Term-document matrix · Document-term matrix · Term frequency · Document frequency · Term frequency-inverse document frequency · Inverse document frequency · Weighting · Term weighting · Document weighting · Log frequency

5.1 Introduction

Following the text preparation and preprocessing, outlined in Chap. 4, the next step is to transform the text documents into a compatible format for text analysis. At this stage, we need to convert the text data into frequencies that can be used in analytic calculations. To build the term-document representation, we borrow some concepts from matrix algebra. In this chapter, before transforming the text data into a term-document matrix representing the frequency of each word in each document, we create an inverted index. Finally, we present several weighting measures that can be used to transform the matrix representation.

5.2 The Inverted Index

The first step toward building a representation of the terms and documents in our document collection is to create an inverted index. An inverted index contains a dictionary of the unique terms or n-grams in the preprocessed tokenized text. The index also contains postings where the documents in which each of the dictionary terms occurs are listed (Manning et al. 2008).

© Springer Nature Switzerland AG 2019
M. Anandarajan et al., *Practical Text Analytics*, Advances in Analytics and Data Science 2, https://doi.org/10.1007/978-3-319-95663-3_5

Table 5.1 Unprocessed and preprocessed text

Documents		
Number	Text	Preprocessed text
1	My favorite dog is fluffy and tan	[favorite] [dog] [fluffy] [tan]
2	The dog is brown and cat is brown	[dog] [brown] [cat] [brown]
3	My favorite hat is brown and coat is pink	[favorite] [hat] [brown] [coat] [pink]
4	My dog has a hat and leash	[dog] [hat] [leash]
5	He has a fluffy coat and brown coats	[fluffy] [coat] [brown] [coat]
6	The dog is brown and fluffy and has a brown coat	[dog] [brown] [fluffy] [brown] [coat]
7	My dog is white with brown spots	[dog] [white] [brown] [spot]
8	The white dog has a pink coat and the brown dog is fluffy	[white] [dog] [pink] [coat] [brown] [dog] [fluffy]
9	The three fluffy dogs and two brown hats are my favorites	[fluffy] [dog] [brown] [hat] [favorite]
10	My fluffy dog has a white coat and hat	[fluffy] [dog] [white] [coat] [hat]

Table 5.2 Inverted index for dcument collection

Dictionary	Postings							
brown	2	3	5	6	7	8	9	
cat	2							
coat	3	5	6	8	10			
dog	1	2	4	6	7	8	9	10
favorite	1	3	9					
fluffy	1	5	6	8	9	10		
hat	3	4	9	10				
leash	4							
pink	3	8						
spot	7							
tan	1							
white	7	8	10					

As described in Chap. 4, we preprocess the text to transform it into a format to create a representation of the term-document information. Table 5.1 displays the sample document collection containing ten documents in which dog owners talk about their dogs. In the text column, the raw document text is displayed. The preprocessed text appears in the third column of the table.

From the preprocessed text, we create an inverted index, illustrated in Table 5.2. The table contains the unique preprocessed terms in the document collection on the left, under Dictionary, and the document numbers in which the terms appear are on the right, under Postings. The terms listed in the dictionary can be considered the terms in the vocabulary. The inverted index creates the foundation for our term frequency representation.

Based on the inverted index, we represent the document collection as a listing of each term-posting pair. This listing includes frequency information or the number of times the term appears in a document. For example, as illustrated in

Table 5.3   Document frequency of the term *brown*

Document	Preprocessed tokenized text	Frequency
1	Favorite dog fluffy tan	0
2	Dog *brown* cat *brown*	2
3	Favorite hat *brown* coat pink	1
4	Dog hat leash	0
5	Fluffy coat *brown* coat	1
6	Dog *brown* fluffy *brown* coat	2
7	Dog white *brown* spot	1
8	White dog pink coat *brown* dog fluffy	1
9	Fluffy dog *brown* hat favorite	1
10	Fluffy dog white coat hat	0

Table 5.4   Term-postings frequency table for the term *brown*

Term	Document	Frequency
brown	1	0
brown	2	2
brown	3	1
brown	4	0
brown	5	1
brown	6	2
brown	7	1
brown	8	1
brown	9	1
brown	10	0

Table 5.3, we create a table by counting the number of times the word *brown* appears in each of the documents.

The term *brown* appears in Documents 2 and 6 twice and appears once in Documents 3, 5, 7, 8, and 9. For the term *brown*, the term-posting pairs and frequency information are displayed in Table 5.4. Now, we can represent our documents containing the term *brown* by their document and frequency.

The remaining 11 terms can be represented in the same way as *brown*. By rearranging the inverted index in Table 5.2 to include the frequency information for the term *brown* from Table 5.3, we have computed the frequency values that will make up the first row of our term-document matrix. Our next step will be to transform this list of frequencies for term-document pairs into a matrix representation for all of the terms and documents in our document collection. We begin by introducing the term-document matrix representation.

5.3   The Term-Document Matrix

Text analysis is made possible using some concepts from matrix algebra. A matrix is a two-dimensional array with m rows and n columns. Matrix A is depicted below. Each entry in the matrix is indexed as a_{ij}, where i represents the row number and j indexes the column number of the entry. There are n columns and m rows in matrix A. a_{11}, for instance, is located in the first row and first column of matrix A.

$$
A = \begin{pmatrix}
a_{11} & a_{12} & \cdots & a_{1n} \\
a_{21} & a_{22} & \cdots & a_{2n} \\
\vdots & \vdots & \ddots & \vdots \\
a_{m1} & a_{m2} & \cdots & a_{mn}
\end{pmatrix}
$$

We model textual information from our document collection in two dimensions, terms and documents. Following text parsing and inverted indexing, we can model the individual terms or tokens in each of the documents in our document collection.

In text mining, we use a specific type of matrix to represent the frequencies of terms in documents. A term-document matrix (TDM) or document-term matrix (DTM) is created to represent a collection of documents for text analysis. In a TDM, the rows correspond to terms, and the columns correspond to documents. Alternatively, in a DTM, the rows correspond to documents, and the columns correspond to terms. An illustration of the setup of the two matrices appears in Fig. 5.1. In the examples of the DTM and TDM layouts in the figure, there are three terms and three documents. The only difference between the two is the placement of the

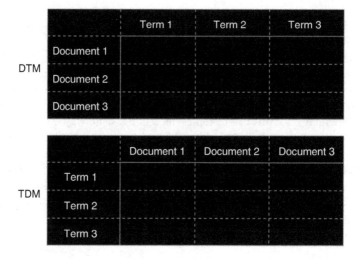

Fig. 5.1 Basic document-term and term-document matrix layouts

Table 5.5 Term-document matrix example

	D1	D2	D3	D4	D5	D6	D7	D8	D9	D10
brown	0	2	1	0	1	2	1	1	1	0
cat	0	1	0	0	0	0	0	0	0	0
coat	0	0	1	0	2	1	0	1	0	1
dog	1	1	0	1	0	1	1	2	1	1
favorite	1	0	1	0	0	0	0	0	1	0
fluffy	1	0	0	0	1	1	0	1	1	1
hat	0	0	1	1	0	0	0	0	1	1
leash	0	0	0	1	0	0	0	0	0	0
pink	0	0	1	0	0	0	0	1	0	0
spot	0	0	0	0	0	0	1	0	0	0
tan	1	0	0	0	0	0	0	0	0	0
white	0	0	0	0	0	0	1	1	0	1

terms and documents, and either can be created and used in text mining analysis. In the example in this chapter, we will build a TDM.

As described, a TDM created to represent a collection of n documents has m rows and n columns, where m represents the total number of terms and n represents the total number of documents. Each entry a_{ij} contains the frequency with which term i occurs in document j. Typically, the number of terms in the TDM will be greater than the number of documents. The unique terms in the preprocessed text column of Table 5.1 are used to create the rows of our TDM. Each document in the table becomes a column in our TDM.

The term *brown* and the other terms in our vocabulary become the rows of our matrix, and the documents will be the columns. The frequency values will include the frequency for each of the term-document pairs. Any term that does not occur in a document will have a value of 0. We represent the document collection introduced in Table 5.1 as a TDM in Table 5.5. In this case, we have a 12-term by 10-document matrix. Note that the row for the term *brown* is the frequency column from Table 5.4. All of the rows in the TDM in Table 5.5 are computed in the same fashion.

Given that a matrix is a collection of points, we can represent the information visually to examine our TDM. In Fig. 5.2, a heat map is used to represent the frequency information in the TDM. The darker colors indicate lower frequency, and the lighter colors indicate higher frequency values in the TDM. The heat map shows that the words *dog* and *brown* are commonly used in our document collection, while *leash* and *spot* are rarely used.

5.4   Term-Document Matrix Frequency Weighting

When creating the TDM in Table 5.5, we used the term frequency values of each term in each document as the values in the matrix. The term frequency of term i in document j is sometimes denoted as $tf_{i,j}$. In using the term frequency in our TDM, the higher the frequency of a given term in a document, the more important

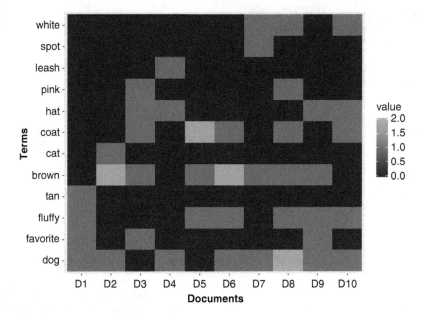

Fig. 5.2 Heat map visualizing the term-document matrix

that term is to the content of that document. For instance, in looking at the TDM in Table 5.5, the term *brown* appears twice in Document 2 and only once in Document 3. This example illuminates a major issue with using term frequency to measure importance. The term *brown* is more important in Document 2 than in Document 3, but is it twice as important? To reduce the impact of a high degree of variability in term frequencies, we can use alternative weighting approaches.

We will explore local weighting, global weighting, and combinatorial weighting approaches. Local weighting measures apply weighting to capture the importance of a term within a specific document in the larger collection of documents. This weighting tells us how much a term contributes to each of the documents. Global weighting is the overall importance of the term in the full collection of documents (Berry et al. 1999). Words that appear frequently, and in many documents, will have a low global weight. Combinatorial weighting combines local and global weighting.

5.4.1 Local Weighting

When applying local weighting, the result will be a matrix with the same dimensions as the original, unweighted TDM. In our case, the result of local weighting will be a 12-word by 10-document-weighted TDM. The local weighting alternatives that we consider are logarithmic (log) frequency and binary/Boolean frequency.

Table 5.6 Log frequency matrix

	D1	D2	D3	D4	D5	D6	D7	D8	D9	D10
brown	0.0	1.1	0.7	0.0	0.7	1.1	0.7	0.7	0.7	0.0
cat	0.0	0.7	0.0	0.0	0.0	0.0	0.0	0.0	0.0	0.0
coat	0.0	0.0	0.7	0.0	1.1	0.7	0.0	0.7	0.0	0.7
dog	0.7	0.7	0.0	0.7	0.0	0.7	0.7	1.1	0.7	0.7
favorite	0.7	0.0	0.7	0.0	0.0	0.0	0.0	0.0	0.7	0.0
fluffy	0.7	0.0	0.0	0.0	0.7	0.7	0.0	0.7	0.7	0.7
hat	0.0	0.0	0.7	0.7	0.0	0.0	0.0	0.0	0.7	0.7
leash	0.0	0.0	0.0	0.7	0.0	0.0	0.0	0.0	0.0	0.0
pink	0.0	0.0	0.7	0.0	0.0	0.0	0.0	0.7	0.0	0.0
spot	0.0	0.0	0.0	0.0	0.0	0.0	0.7	0.0	0.0	0.0
tan	0.7	0.0	0.0	0.0	0.0	0.0	0.0	0.0	0.0	0.0
white	0.0	0.0	0.0	0.0	0.0	0.0	0.7	0.7	0.0	0.7

5.4.1.1   Logarithmic (Log) Frequency

Log frequency is a weighting method that reduces the effect of large differences in frequencies (Dumais 1991). The base of the logarithm can vary. Below, the natural logarithm, denoted ln, is used. Table 5.6 illustrates the log frequency-weighted TDM. As the table shows, the weight of the terms appearing twice in a document has a value of 1.1, and terms appearing once in a document have a value of 0.7. This method reduces the difference between the two weights from 1 to 0.4. The log frequency of term i in document j, $lf_{i,j}$, is calculated as

$$lf_{i,j} = \begin{cases} \ln\left(tf_{i,j} + 1\right), & \text{if } tf_{i,j} > 0 \\ 0, & \text{otherwise} \end{cases}.$$

5.4.1.2   Binary/Boolean Frequency

Binary frequency captures whether a word appears in a document, without consideration of how many times it appears. In the binary frequency-weighted TDM in Table 5.7, there is no difference between a term occurring once or twice in a document. This approach is equivalent to recording if a term appears in a document. A binary frequency matrix can be used to perform further weighting on the TDM. The binary frequency of term i in document j, $n_{i,j}$, is calculated as

$$n_{i,j} = \begin{cases} 1, & \text{if } tf_{i,j} > 0 \\ 0, & \text{otherwise} \end{cases}.$$

Table 5.7 Binary frequency matrix

	D1	D2	D3	D4	D5	D6	D7	D8	D9	D10
brown	0	1	1	0	1	1	1	1	1	0
cat	0	1	0	0	0	0	0	0	0	0
coat	0	0	1	0	1	1	0	1	0	1
dog	1	1	0	1	0	1	1	1	1	1
favorite	1	0	1	0	0	0	0	0	1	0
fluffy	1	0	0	0	1	1	0	1	1	1
hat	0	0	1	1	0	0	0	0	1	1
leash	0	0	0	1	0	0	0	0	0	0
pink	0	0	1	0	0	0	0	1	0	0
spot	0	0	0	0	0	0	1	0	0	0
tan	1	0	0	0	0	0	0	0	0	0
white	0	0	0	0	0	0	1	1	0	1

5.4.2 Global Weighting

Global weighting indicates the importance of a term in the whole document collection, rather than in individual documents. When applying global weighting, the result of the calculations will be a vector of values that is the length of the total number of terms, which in our case is 12. The global weighting alternatives that we consider are document frequency, global frequency, and inverse document frequency.

5.4.2.1 Document Frequency (*df*)

Document frequency can be derived using the binary frequency-weighted TDM by summing the rows of the binary frequency-weighted TDM. Document frequency, df_i, is calculated as

$$df_i = \sum_{j=1}^{D} n_{i,j},$$

where $n_{i,j}$ is the binary frequency-weighted matrix and D is the total number of documents.

As an example, we calculate the document frequency of the word *brown* by adding the values in the row for *brown* in the binary frequency-weighted TDM.

$$df_{brown} = \sum_{j=1}^{10} n_{brown,j} = 0+1+1+0+1+1+1+1+1+0 = 7.$$

We find that the document frequency of the term *brown* is 7, meaning that the term *brown* appears in seven out of the ten documents in the collection. The document frequency values for each of the 12 terms can be calculated the same way and are plotted in Fig. 5.3.

5.4.2.2   Global Frequency (*gf*)

Global frequency measures the frequency of terms across all documents and is calculated as

$$gf_i = \sum_{j=1}^{D} tf_{i,j},$$

where $tf_{i,j}$ is the frequency of term i in document j and D is the number of documents.

As an example, we compute the global frequency of the word *brown*. Using the formula, we calculate

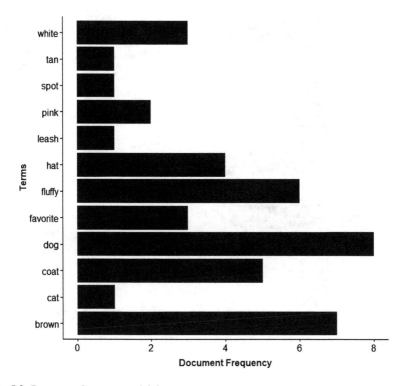

Fig. 5.3 Document frequency weighting

$$gf_{brown} = \sum_{j=1}^{10} tf_{brown,j} = 0+2+1+0+1+2+1+1+1+0 = 9.$$

The global frequencies of the other terms are computed in the same way and are displayed in Fig. 5.4.

5.4.2.3   Inverse Document Frequency (*idf*)

In inverse document frequency, rare terms have higher weights, and frequent terms have lower weights (Dumais 1991). Inverse document frequency, idf_i, is calculated as

$$idf_i = \log_2\left(\frac{n}{df_i}\right) + 1,$$

where n is the total number of documents in the collection.

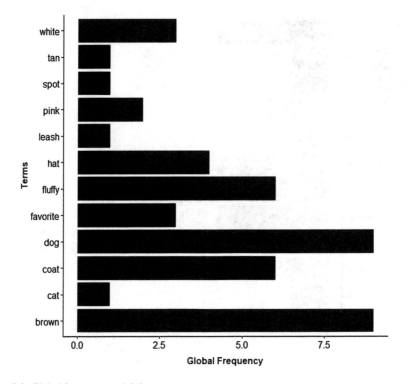

Fig. 5.4 Global frequency weighting

To find the inverse document frequency of the word *brown*, we would calculate it as follows

$$idf_{brown} = \log_2\left(\frac{10}{df_{brown}}\right) + 1 = \log_2\left(\frac{10}{7}\right) + 1 = 1.51.$$

The *idf* for each of the words can be computed in the same way. Figure 5.5 depicts the results of calculating the inverse document frequency for each of the terms.

5.4.3   Combinatorial Weighting: Local and Global Weighting

Combinatorial weighting combines local and global frequency weighting to consider the importance of each of the terms in the documents individually and in the document collection. In some cases, combinatorial weighting can also include normalization based on the total number of terms in each document. Here, we will focus on one of the most common combinatorial weighting measures, term frequency-inverse document frequency.

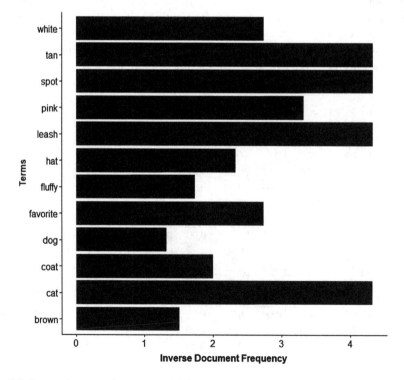

Fig. 5.5 Inverse document frequency weighting

5.4.3.1   Term Frequency-Inverse Document Frequency (*tfidf*)

Term frequency-inverse document frequency weighting combines term frequency
and inverse document frequency by multiplying the local term frequency weight by
the global inverse document frequency weight. *tfidf* is calculated as

$$tfidf_{i,j} = tf_{i,j} * idf_i,$$

where $tf_{i,j}$ is term frequency and idf_i is inverse document frequency.

   tfidf is high when a term occurs many times in a few documents and is low when
a term occurs in all, most or many documents. Intuitively, if a word appears frequently in a document or the collection of documents, it would make sense to consider the term to be important. However, the more frequently a term appears across
documents, the less it actually helps with understanding the textual content.

   The use of this method "balances the importance of a term to a document by its
frequency in that document, evidenced by its frequency in that document, against
a term's overall discriminative ability, based on its distribution across the collection as a whole" (Jessup and Martin 2001, p. 5). The *tfidf*-weighted TDM matrix
appears in Table 5.8. As the table shows, terms such as *spots*, *cat*, and *tan*, which
appear infrequently across the collection of documents but appear frequently in a
particular document, have a high *tfidf* weight value in the documents in which they
appear. The word *dog*, which appears frequently in the collection of documents,
has a low weighting in the documents in which it occurs because of its high global
frequency. The *tfidf*-weighted TDM, which is found by multiplying the term
frequency-weighted TDM and the inverse document frequency vector, is shown in
Table 5.8.

Table 5.8   *tfidf*-weighted TDM

	D1	D2	D3	D4	D5	D6	D7	D8	D9	D10
brown	0.00	3.03	1.51	0.00	1.51	3.03	1.51	1.51	1.51	0.00
cat	0.00	4.32	0.00	0.00	0.00	0.00	0.00	0.00	0.00	0.00
coat	0.00	0.00	2.00	0.00	4.00	2.00	0.00	2.00	0.00	2.00
dog	1.32	1.32	0.00	1.32	0.00	1.32	1.32	2.64	1.32	1.32
favorite	2.74	0.00	2.74	0.00	0.00	0.00	0.00	0.00	2.74	0.00
fluffy	1.74	0.00	0.00	0.00	1.74	1.74	0.00	1.74	1.74	1.74
hat	0.00	0.00	2.32	2.32	0.00	0.00	0.00	0.00	2.32	2.32
leash	0.00	0.00	0.00	4.32	0.00	0.00	0.00	0.00	0.00	0.00
pink	0.00	0.00	3.32	0.00	0.00	0.00	0.00	3.32	0.00	0.00
spot	0.00	0.00	0.00	0.00	0.00	0.00	4.32	0.00	0.00	0.00
tan	4.32	0.00	0.00	0.00	0.00	0.00	0.00	0.00	0.00	0.00
white	0.00	0.00	0.00	0.00	0.00	0.00	2.74	2.74	0.00	2.74

5.5 Decision-Making

Having reviewed the frequency-weighting options, it is probably clear that each weighting schema has its own strengths and weaknesses. The choice of weighting method will depend on both the data and the intended modeling and analysis method. For instance, the first analysis method that we will explore, latent semantic analysis (LSA), covered in Chap. 6, is well suited to *tfidf* weighting. On the other hand, some topic models, including latent Dirichlet allocation (LDA), which is covered in Chap. 8, require the unweighted TDM as an input for the analysis.

In addition to the modeling and analysis considerations that influence the choice of weighting, there are some inherent weaknesses in raw frequency data that encourage the use of weighted TDMs. First, longer documents will have higher term counts than shorter documents. Additionally, high-frequency terms may be less important than lower-frequency terms. Indeed, the idea of a stop word list is based on the notion that the most common terms in a language will be the lowest in content. Terms such as *the* and *an* may be high in frequency in a document collection but will certainly be low in content.

Key Takeaways
- An inverted index represents term-posting frequency information for a document collection.
- A term-document or document-term matrix representation transforms pre-processed text data into a matrix representation that can be used in analysis.
- Local, global, and combinatorial weighting can be applied to the term-document or document-term matrix.

References

Berry, M. W., Drmac, Z., & Jessup, E. R. (1999). Matrices, vector spaces, and information retrieval. *SIAM Review, 41*(2), 335–362.

Dumais, S. T. (1991). Improving the retrieval of information from external sources. *Behavior Research Methods, Instruments, & Computers, 23*(2), 229–236.

Jessup, E. R., & Martin, J. H. (2001). Taking a new look at the latent semantic analysis approach to information retrieval. *Computational Information Retrieval, 2001*, 121–144.

Manning, C., Raghavan, P., & Schütze, H. (2008). *Introduction to information retrieval*. Cambridge: Cambridge University Press. https://doi.org/10.1017/CBO9780511809071.

Further Reading

For more about the term-document representation of text data, see Berry et al. (1999) and Manning et al. (2008).

Part III
Text Analysis Techniques

Chapter 6
Semantic Space Representation and Latent Semantic Analysis

Abstract In this chapter, we introduce latent semantic analysis (LSA), which uses singular value decomposition (SVD) to reduce the dimensionality of the document-term representation. This method reduces the large matrix to an approximation that is made up of fewer latent dimensions that can be interpreted by the analyst. Two important concepts in LSA, cosine similarity and queries, are explained. Finally, we discuss decision-making in LSA.

Keywords Latent semantic analysis (LSA) · Singular value decomposition (SVD) · Latent semantic indexing (LSI) · Cosine similarity · Queries

6.1 Introduction

In Chapter 5, we built a term-document matrix (TDM) based on the text in our document collection. This matrix-based representation allows us to consider documents as existing in term space and terms as existing in document space. In this chapter, we present the latent semantic analysis (LSA) of the TDM. LSA is a fully automatic semantic space modeling approach in which terms are points in high-dimensional space and the spatial closeness between those points represents their semantic association (Landauer and Dumais 1997). Semantic representation tries to reveal meaning that can be hidden in the documents. Semantic knowledge extends beyond meaning to consider relations among terms and the hidden meaning and concepts present in the documents.

The examples in this chapter use the *tfidf*-weighted TDM created in Chap. 5, which includes 12 terms and 10 documents in which dog owners describe their dogs. *Tfidf* weighting is a popular weighting method used in LSA, because it combines a local and global weighting function to dampen the impact of high-frequency terms and give more weight to less frequently occurring documents that occur in fewer documents. The *tfidf*-weighted matrix used in this chapter is presented in Table 5.8. We use a weighted TDM because it produces improved results over models built with no weighting (Dumais 1991). We begin by plotting relationships based on the simple, unweighted TDM to conceptualize the term and document spaces.

© Springer Nature Switzerland AG 2019

M. Anandarajan et al., *Practical Text Analytics*, Advances in Analytics and Data Science 2, https://doi.org/10.1007/978-3-319-95663-3_6

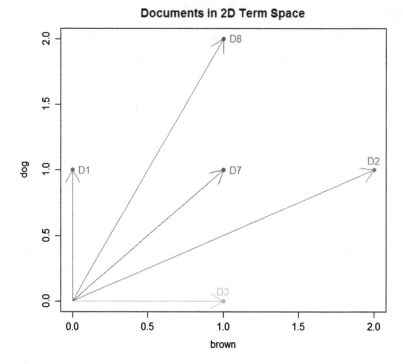

Fig. 6.1 Two-dimensional representation of the first five documents in term space for the terms *brown* and *dog*

To begin, we visualize a vector space representation. As shown in Fig. 6.1, we plot the raw frequency information for the words *brown* and *dog* using Documents 1, 2, 3, 7, and 8. Plotting documents in term space can help us understand the distance between documents and provides a geometric representation of our TDM. This figure depicts these five documents as vectors in two-dimensional term space. As shown, the documents are the points in this space. For instance, Document 1 is a point located at (0,1) in the *brown* and *dog* space, because *brown* does not occur in the document and *dog* occurs once. Document 2 is a point located at (2,1) in the *brown* and *dog* space because *brown* occurs twice and *dog* occurs once in the document. The angles formed by vectors in spatial representations are the basis for an important measure of association, cosine similarity, which will be covered in Sect. 6.2.

For more than two terms, we can visualize documents in term space in higher dimensions, as shown in Fig. 6.2. The three-dimensional term space in that figure includes the terms *brown*, *coat*, and *favorite*. In this figure, we visualize the frequencies of each of the three terms in each of our documents in our document collection. The same plots can be created to represent terms in document space. Due to the number of dimensions in the TDM, we are limited in how we can visualize these associations. However, the use of the semantic space representation allows us to model these associations in much higher dimensions than our graph allows.

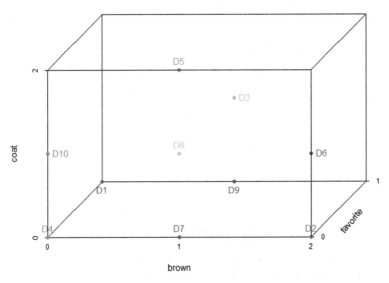

Fig. 6.2 Three-dimensional representation of the ten documents in term space for the terms *brown, coat* and *favorite*

6.2   Latent Semantic Analysis (LSA)

Dumais et al. (1988) and Deerwester et al. (1990) first introduced latent semantic analysis (LSA) as latent semantic indexing (LSI), due to its objective of indexing text. LSA extends the concept to all analytic applications beyond indexing. LSA creates a vector space representation of our original matrix using singular value decomposition (SVD). Specifically, LSA is an application of SVD to identify latent meaning in the documents through dimension reduction. The original TDM is assumed to be too big and sparse to be useful and/or meaningful. In real-world text applications, the TDM representation of a document collection can be very large and difficult to interpret or understand. LSA not only reduces the dimensionality but also identifies latent dimensions based on singular values. In order to understand how LSA works, we first need to familiarize ourselves with SVD.

6.2.1   Singular Value Decomposition (SVD)

LSA relies on SVD to identify latent information in the TDM. SVD splits a matrix, in our case the TDM, into three smaller matrices that, when multiplied, are equivalent to the original matrix. After this decomposition, we reduce the size of our three component matrices further by choosing to keep a smaller number of dimensions.

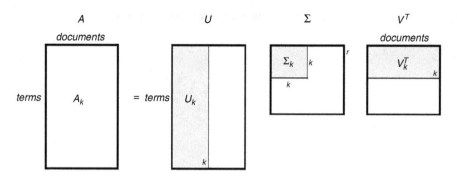

Fig. 6.3 SVD process in LSA, based on Martin and Berry (2007)

SVD is used in semantic space modeling to create smaller approximations of large document-term matrices. The truncated matrix created through SVD has four important purposes: latent meaning, noise reduction, high-order co-occurrence, and sparsity reduction (Turney and Pantel 2010, pp. 159–160).

In SVD, we calculate $A = U\Sigma V^T$, where A is the term-document matrix, U is the left singular vector of words, Σ is a matrix with weight values on the diagonal, and V is the right singular vector of documents. r is the rank of the matrix A. If we reduce r to a smaller number, k, we create an approximation of the original matrix. r can then be reduced to k, where A_k is then a lower dimensional, rank k approximation of the original matrix A. In addition, the dimensions of the three component matrices are adjusted from r to k. SVD can be applied to any rectangular matrix to decompose a larger matrix into the product of three smaller matrices. Figure 6.3 depicts the SVD of the A matrix.

6.2.2   LSA Example

When performing SVD, the TDM is known as the A matrix, which is the matrix of which we want to create a three-component representation. The rank of our A matrix, r, is 10, which is the minimum of the number of rows and number of columns in our A matrix. More formally, rank is

$$r = \min \begin{cases} m \\ n \end{cases}$$

where m is the total number of terms and n is the total number of documents.

If the number of documents in the TDM is larger than the number of terms in our TDM, the rank will be equal to the number of terms. On the other hand, if the number of terms is larger, the rank will be equal to the number of documents. In our case, we have the latter situation, because our terms outnumber our documents.

$A =$

	D1	D2	D3	D4	D5	D6	D7	D8	D9	D10
Brown	0.0	3.0	1.5	0.0	1.5	3.0	1.5	1.5	1.5	0.0
Cat	0.0	4.3	0.0	0.0	0.0	0.0	0.0	0.0	0.0	0.0
Coat	0.0	0.0	2.0	0.0	4.0	2.0	0.0	2.0	0.0	2.0
Dog	1.3	1.3	0.0	1.3	0.0	1.3	1.3	2.6	1.3	1.3
Favorite	2.7	0.0	2.7	0.0	0.0	0.0	0.0	0.0	2.7	0.0
Fluffy	1.7	0.0	0.0	0.0	1.7	1.7	0.0	1.7	1.7	1.7
Hat	0.0	0.0	2.3	2.3	0.0	0.0	0.0	0.0	2.3	2.3
Leash	0.0	0.0	0.0	4.3	0.0	0.0	0.0	0.0	0.0	0.0
Pink	0.0	0.0	3.3	0.0	0.0	0.0	0.0	3.3	0.0	0.0
Spot	0.0	0.0	0.0	0.0	0.0	0.0	4.3	0.0	0.0	0.0
Tan	4.3	0.0	0.0	0.0	0.0	0.0	0.0	0.0	0.0	0.0
White	0.0	0.0	0.0	0.0	0.0	0.0	2.7	2.7	0.00	2.7

The U matrix is our term matrix and the left singular vector. The U matrix has 12 rows, a row for each of the terms in the TDM. The number of columns in the U matrix is ten, because the number of columns is equal to the rank of the A matrix.

$U =$

−0.4	−0.3	−0.4	0.1	0.2	−0.1	0.3	0.2	−0.3	0.4
−0.1	−0.2	−0.5	0.3	0.3	−0.1	−0.2	−0.2	0.5	−0.3
−0.4	−0.1	0.0	−0.6	0.0	0.3	0.2	0.2	0.4	−0.1
−0.4	−0.1	0.1	0.3	0.0	0.2	−0.3	−0.1	−0.4	−0.1
−0.3	0.6	−0.1	0.1	−0.1	−0.3	0.3	0.0	0.0	−0.4
−0.4	0.1	0.0	−0.1	−0.2	0.4	0.0	−0.2	−0.3	−0.4
−0.3	0.3	0.3	0.1	0.4	−0.1	0.3	−0.4	0.2	0.4
−0.1	0.2	0.3	0.3	0.6	0.3	−0.1	0.5	0.0	−0.2
−0.3	0.1	0.1	−0.3	0.0	−0.6	−0.5	0.3	0.0	0.0
−0.1	−0.4	0.3	0.4	−0.3	−0.2	0.5	0.3	0.2	−0.2
−0.1	0.4	−0.2	0.3	−0.5	0.2	−0.2	0.3	0.3	0.4
−0.3	−0.3	0.4	0.1	−0.3	0.0	−0.2	−0.3	0.2	0.1

The Σ matrix contains the singular values on the diagonal, and the rest of the matrix is zeroes. The Σ matrix is a symmetric, or square matrix, with the number of rows and columns equal to the rank of our A matrix. For this reason, our Σ matrix has ten rows and ten columns. The singular values on the diagonal of the matrix are in decreasing order. The largest singular value is in the first column, and the smallest singular value is in the last column. This will be the case in any SVD application.

$\Sigma =$

9.9	0.0	0.0	0.0	0.0	0.0	0.0	0.0	0.0	0.0
0.0	6.0	0.0	0.0	0.0	0.0	0.0	0.0	0.0	0.0
0.0	0.0	5.4	0.0	0.0	0.0	0.0	0.0	0.0	0.0
0.0	0.0	0.0	5.3	0.0	0.0	0.0	0.0	0.0	0.0
0.0	0.0	0.0	0.0	5.1	0.0	0.0	0.0	0.0	0.0
0.0	0.0	0.0	0.0	0.0	4.2	0.0	0.0	0.0	0.0
0.0	0.0	0.0	0.0	0.0	0.0	3.5	0.0	0.0	0.0
0.0	0.0	0.0	0.0	0.0	0.0	0.0	3.0	0.0	0.0
0.0	0.0	0.0	0.0	0.0	0.0	0.0	0.0	2.3	0.0
0.0	0.0	0.0	0.0	0.0	0.0	0.0	0.0	0.0	0.7

The V^T matrix, in which T represents the transpose, is the document matrix and the right singular vector. The number of rows in the V^T matrix is equal to the rank of our A matrix and is ten. The number of columns in our V^T matrix is equal to the number of documents, ten. When the number of terms is larger than the number of documents in a TDM, V^T will have the same number of rows and columns, because the rank is equal to the number of documents. On the other hand, if the number of documents is larger than the number of terms in a TDM, the U matrix will have the same number of rows and columns, because the rank value, r, will be the number of terms.

$V^T =$

−0.2	−0.2	−0.4	−0.1	−0.3	−0.3	−0.2	−0.5	−0.3	−0.3
0.6	−0.3	0.3	0.2	−0.1	−0.1	−0.5	−0.2	0.3	0.0
−0.2	−0.7	0.0	0.4	−0.1	−0.2	0.3	0.2	−0.1	0.3
0.4	0.4	−0.3	0.3	−0.4	−0.1	0.5	−0.2	0.2	−0.1
−0.5	0.4	0.2	0.7	0.0	0.0	−0.3	−0.1	0.1	0.0
0.2	−0.1	−0.7	0.3	0.4	0.3	−0.2	−0.1	−0.1	0.3
−0.2	−0.2	0.2	−0.1	0.3	0.2	0.4	−0.7	0.4	0.0
0.2	−0.2	0.2	0.3	0.3	0.2	0.2	0.1	−0.4	−0.7
0.2	0.3	0.3	0.0	0.3	−0.5	0.1	−0.3	−0.5	0.4
0.1	0.0	0.2	0.0	−0.5	0.6	0.0	−0.3	−0.4	0.3

Now that we have used SVD to create the three-component approximation of the original TDM, we can create a lower-rank approximation of A to reduce the size. The Σ matrix has ten singular values along the diagonal, equal to the rank of our A matrix. We want to reduce the number of singular values, thereby reducing the size of each of our three component matrices, because each of them has at least one dimension that depends on the rank of A. We choose a number, k, and reduce the size of each of our component matrices' dimensions from r to k.

We choose to retain $k = 3$ singular vectors or three latent dimensions. The reduced U, Σ, and V^T matrices are shown below. Setting $k = 3$, the U matrix has 12 rows and 3 columns, the Σ matrix has 3 rows and 3 columns, and the V^T matrix has 3 rows and 10 columns.

$U =$

−0.4	−0.3	−0.4
−0.1	−0.2	−0.5
−0.4	−0.1	0.0
−0.4	−0.1	0.1
−0.3	0.6	−0.1
−0.4	0.1	0.0
−0.3	0.3	0.3
−0.1	0.2	0.3
−0.3	0.1	0.1
−0.1	−0.4	0.3
−0.1	0.4	−0.2
−0.3	−0.3	0.4

$\Sigma =$

9.9	0.0	0.0
0.0	6.0	0.0
0.0	0.0	5.4

$V^T =$

−0.2	−0.2	−0.4	−0.1	−0.3	−0.3	−0.2	−0.5	−0.3	−0.3
0.6	−0.3	0.3	0.2	−0.1	−0.1	−0.5	−0.2	0.3	0.0
−0.2	−0.7	0.0	0.4	−0.1	−0.2	0.3	0.2	−0.1	0.3

After choosing k, we multiply the above component matrices to find A_k, our reduced rank approximation of the original A matrix. Our A_k matrix, A_3, appears in Table 6.1 in what is referred to as the LSA space.

Table 6.1 The LSA space

	D1	D2	D3	D4	D5	D6	D7	D8	D9	D10
brown	0.6	3.0	1.2	−0.7	1.8	2.3	1.1	2.1	1.0	0.7
cat	0.1	2.6	0.0	−1.4	0.8	1.2	0.0	0.3	0.1	−0.6
coat	0.7	1.1	1.5	0.5	1.4	1.5	1.3	2.3	1.2	1.5
dog	0.6	0.8	1.3	0.6	1.1	1.2	1.2	2.0	1.0	1.4
favorite	2.6	0.1	2.0	0.8	0.6	0.6	−1.2	0.4	2.0	0.5
fluffy	1.3	0.7	1.6	0.6	1.0	1.1	0.4	1.5	1.4	1.1
hat	1.4	−1.1	1.7	1.5	0.4	0.2	0.2	1.2	1.4	1.4
leash	0.3	−1.4	0.6	1.1	−0.2	−0.4	0.2	0.4	0.4	0.8
pink	0.7	0.2	1.2	0.7	0.8	0.8	0.8	1.5	1.0	1.2
spot	−1.3	0.0	−0.2	0.2	0.3	0.3	1.7	1.2	−0.5	0.9
tan	1.8	0.1	1.1	0.3	0.2	0.2	−1.3	−0.2	1.2	−0.1
white	−1.0	−0.2	0.5	0.9	0.8	0.7	2.4	2.3	0.1	1.9

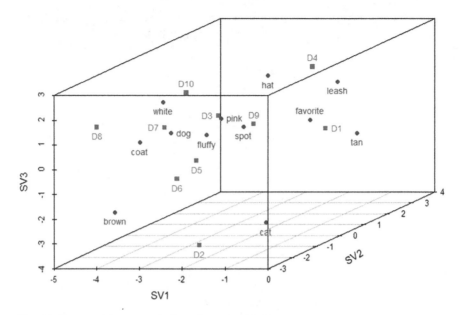

Fig. 6.4 Terms and documents in three-dimensional LSA vector space

Figure 6.4 shows the documents and terms in three-dimensional LSA vector space. The documents are depicted as light blue squares, and the terms are darker blue dots. We created this representation by multiplying our reduced U and V^T matrices by the Σ matrix. This visualization has a clear advantage over the visualization in Fig. 6.2 in that we can view both documents and terms concurrently across the three dimensions. Each of the documents and terms in the LSA vector space can be thought of as vectors emanating from the origin.

For concreteness, Fig. 6.5 presents a two-dimensional version of Fig. 2.4 with only the first two dimensions. The dashed lines drawn from the origin to *leash* and *cat* can be drawn for each document and term, because they are vectors. Using these concepts, we can map the associations between terms, documents, and terms and documents. This depiction gives rise to a geometric measure of closeness known as cosine similarity.

6.3 Cosine Similarity

The primary method of measuring the association between terms and documents in LSA is cosine similarity. Cosine similarity is a means of measuring the semantic similarity of words, regardless of their actual co-occurrence in text documents (Landauer and Dumais 1997). By applying LSA, we model the terms and documents in our TDM in vector space. Doing so gives us the ability to model many right triangles emanating from the origin to the documents and terms. While the

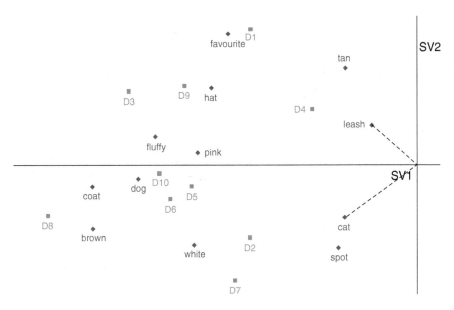

Fig. 6.5 Terms and documents of a two-dimensional LSA solution across the first two dimensions

calculation is made in multidimensional space, we consider cosine similarity in two dimensions for concreteness.

From trigonometry, we know that the cosine is the angle between two vectors. In Fig. 6.5, an angle is formed by the two term vectors meeting at the origin. The rows of the U matrix are converted to column vectors for the calculation. Cosine similarity can be applied to terms, documents, or both. It can also be applied to queries, or pseudo-documents, which will be covered in Sect. 6.5. Cosine similarity scores range between -1 and 1 but are typically greater than 0. The cosine similarity for two terms, t_1 and t_2, is calculated as.

$$Cosine(t_1, t_2) = \frac{t_1^T t_2}{t_1 t_2},$$

where t_1^T is the transpose of the t_1 vector and $\|\|$ represents the vector norm. In the case of documents, $Cosine(d_1, d_2)$, d_1 and d_2 replace t_1 and t_2, representing Document 1 and Document 2, respectively.

Our example uses the formula above to estimate the cosine similarity for two terms, rounding to two decimal places for all intermediate calculations. To calculate the cosine similarity measurement between the terms *fluffy* and *pink*, we use the row vectors from the LSA space, A_3 corresponding to these words.

fluffy:									
1.25	0.74	1.58	0.56	1.03	1.12	0.43	1.54	1.35	1.10

pink:

0.71	0.17	1.24	0.74	0.78	0.78	0.75	1.50	0.98	1.22

$$fluffy^T pink = \sum_{i=1}^{10} fluffy_i * pink_i$$

$$fluffy^T pink = 0.89 + 0.13 + 1.96 + 0.41 + 0.80 + 0.87 + 0.32 + 2.31 + 1.32 + 1.34 = 10.35$$

The numerator is the cross product, or the dot product, of the two vectors, which is calculated to be 10.35. The denominator is the product of the matrix norms of the two vectors, which can be computed as the sum of the squared vector values.

$$fluffy = \sqrt{1.25^2 + 0.74^2 + 1.58^2 + 0.56^2 + 1.03^2 + 1.12^2 + 0.43^2 + 1.54^2 + 1.35^2 + 1.10^2} = 3.58$$

$$pink = \sqrt{0.71^2 + 0.17^2 + 1.24^2 + 0.74^2 + 0.78^2 + 0.78^2 + 0.75^2 + 1.50^2 + 0.98^2 + 1.22^2} = 3.02$$

$$fluffypink = 10.81$$

The cosine similarity measure of the terms *fluffy* and *pink* can then be calculated by $Cosine(fluffy,pink) = \frac{10.35}{10.81} = 0.96$. The cosine similarity values for *fluffy* and the remaining 11 words are calculated in the same way. Table 6.2 presents the cosine similarity values for the term *fluffy* in descending order. The terms *pink*, *dog*, and *coat* have high cosine similarity values with the word *fluffy*. On the other hand, *spots*, *cat*, and *leash* have low cosine values with the word *fluffy*.

The cosine similarity values for all term-term relationships are displayed in Table 6.3. Pairs of terms with the highest cosine similarity values around 1.0 are *coat* and *dog*, *coat* and *pink*, *dog* and *pink*, *favorite* and *tan*, and *fluffy* and *pink*. Pairs of terms with very high cosine similarity values around 0.90 are *brown* and *coat*,

Table 6.2 Cosine similarity measures for *fluffy*, in descending order

Term	Cosine
pink	0.96
dog	0.94
coat	0.94
hat	0.79
brown	0.79
favorite	0.75
white	0.55
tan	0.54
leash	0.32
cat	0.27
spots	0.19

Table 6.3 Term-term cosine similarity measures

	brown	cat	coat	dog	favorite	fluffy	hat	leash	pink	spot	tan	white
brown	0.0											
cat	0.8	0.0										
coat	0.9	0.3	0.0									
dog	0.8	0.3	1.0	0.0								
favorite	0.3	0.0	0.5	0.5	0.0							
fluffy	0.8	0.3	0.9	0.9	0.8	0.0						
hat	0.3	−0.4	0.7	0.7	0.8	0.8	0.0					
leash	−0.3	−0.8	0.2	0.3	0.4	0.3	0.8	0.0				
pink	0.7	0.1	1.0	1.0	0.6	1.0	0.9	0.5	0.0			
spots	0.3	0.0	0.5	0.5	−0.5	0.2	0.1	0.2	0.4	0.0		
tan	0.2	0.0	0.2	0.2	1.0	0.5	0.6	0.2	0.4	−0.7	0.0	
white	0.5	−0.1	0.8	0.8	−0.1	0.5	0.5	0.4	0.7	0.9	−0.4	0.0

fluffy and *coat*, *fluffy* and *dog*, *hat* and *pink*, and *spots* and *white*. The lowest cosine similarity value is for the terms *cat* and *leash*. This result seems reasonable, because these terms should be unrelated.

6.4   Queries in LSA

In the field of information retrieval (IR), the use of the LSA space to explore queries is an essential tool. Anytime you open your browser to a search engine and type in search keywords, you are using a query. Based on the keywords that you provide, the search engine returns websites that it believes match your search criteria. In a similar way, LSA uses the cosine measures to find documents that are similar to words that you designate as query terms (Deerwester et al. 1990). A query is represented as a scaled, weighted sum of the component term vectors. A query is equal to

$$query = q^T U_k \Sigma_k^{-1},$$

where q^T is a vector of the terms in the query, U_k is the term matrix, and Σ_k^{-1} is the inverse of the Σ_k matrix. Multiplying by the inverse of a matrix is equivalent to dividing by the matrix that is inverted. The query is a pseudo-document with a vector representation, which can be compared to the documents in the collection.

For instance, a query could include the component terms *tan*, *brown*, and *pink*. The q^T vector of this query is

$q^T = [1\ 0\ 0\ 0\ 0\ 0\ 0\ 0\ 1\ 0\ 1\ 0]$, based on the pseudo query below.[1]

brown	cat	coat	dog	favorite	fluffy	hat	leash	pink	spot	tan	white
1	0	0	0	0	0	0	0	1	0	1	0

---

[1] Note: The q^T vector is created using binary frequency, because at this stage weighting cannot be calculated and applied to the pseudo-document.

Table 6.4 Cosine values between the query (*brown*, *pink*, *tan*) and documents in descending order by cosine similarity value

Document	Cosine
6	0.81
5	0.78
9	0.73
2	0.71
1	0.69
3	0.66
8	0.24
10	−0.08
7	−0.30
4	−0.30

Using the query formula above, we find that the query is equal to (−0.08, 0.03, −0.09). We use this result to determine the cosine similarity values for the query and each of the documents in the document collection to find the documents most associated with the query. Table 6.4 displays the list of documents in descending order of similarity. As the table illustrates, Documents 6 and 5 are most closely associated with the query, while 7 and 4 are the least associated with, or similar, to the query.

This similarity is based on the angles between the vectors, not physical proximity. Figure 6.6 shows the documents across the three dimensions. The origin is denoted (0, 0, 0), and the pseudo-document query is labeled "query." A thick black line is drawn between the origin and the query. Since Documents 6 and 5 have the highest cosine similarity, these vectors are drawn in green. The document vectors with the lowest cosine similarity, 4 and 7, are plotted in red. As shown, this line between Document 6 and the query nearly overlaps, resulting in a very small angle between the two vectors and the highest cosine similarity. On the other hand, while Document 7 is physically close to the query, the angle between the two vectors is much larger than between the query and Document 6.

6.5   Decision-Making: Choosing the Number of Dimensions

The choice of the number of singular values is an important decision in LSA modeling. If too few SVD dimensions are retained, we run the risk of losing important information. On the other hand, if we keep too many, our calculations and solution may be too complex to be meaningful. For this reason, simple solutions are preferable. Past research has suggested that in the case of big data and very large TDMs (or DTMs), using between 100 and 300 dimensions provides good performance (Berry and Browne 1999; Dumais 1991; Landauer and Dumais 1997).

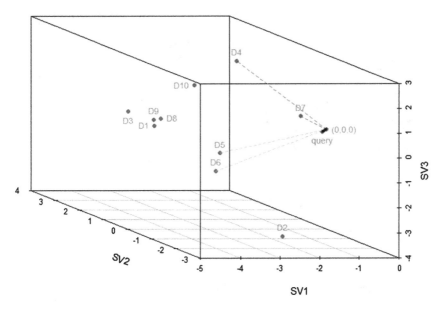

Fig. 6.6 Rotated plot of the query and Document 6 vectors in three-dimensional LSA vector space

In Fig. 6.7, known as a scree plot, we plot the number of singular vectors and the amount of variance explained. A scree plot is a visual aid used to evaluate the tradeoff between efficiency and complexity. In doing so, we look for an elbow in the graph, or the point at which the slope between points is very small, to determine the number of singular vectors to keep. As is evident in the figure, the identification of the elbows is a subjective decision, as is the number of singular values to retain. In the scree plot below, it appears that three dimensions are appropriate.

LSA has many benefits and is widely used in information retrieval and text mining applications. There are many advantages of using LSA, including its ability to handle the sparsity, size, and noise associated with a TDM. In the case of large document collections, the associated TDM will be large and sparse. In addition, due to the uncertainty involved and the qualitative nature of the text data, the data are noisy. LSA can cut through a lot of the noise of the TDM because it is rooted in dimension reduction. Additionally, in reducing the dimensionality, it uncovers latent factors that are otherwise hidden within the data. In LSA, indirect co-occurrences, where, for instance, two words are related through a third word, become important. LSA allows us to compute an association measure, cosine similarity, on a lower-rank matrix rather than our original TDM.

LSA has the innate ability to uncover deep semantic relationships in the terms and documents in the space (Landauer et al. 1998). It can identify similarity that stretches beyond just synonymy and is able to determine the importance of terms (Hu and Liu 2004). LSA handles the types of noisy, sparse matrices that are produced in text analysis through the use of SVD. Additionally, it can create pseudo-documents to measure similarity between existing documents and queries. While

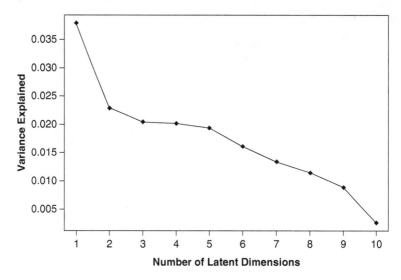

Fig. 6.7 Scree plot showing variance explained by number of singular vectors

the traditional LSA methods do not account for word ordering, because they assume a bag-of-words representation, newer methods extend the methods to incorporate word ordering. LSA can also be applied to TDMs that are based on n-grams or tokens larger than $n = 1$.

Despite the many benefits of LSA, there are limitations to its application. An LSA space is created for a particular document collection, and the results depend heavily on the type of weighting chosen and the number of singular vectors or latent factors retained. The decisions made by the analyst are particularly impactful, and in this sense, the analysis is both an art and a science.

Key Takeaways
- Latent semantic analysis (LSA) can uncover underlying or latent meaning in text.
- LSA uses singular value decomposition (SVD) to reduce the dimensionality of the TDM.
- Cosine similarity based on the LSA space can be used to assess the closeness of term-term and document-document relationships.
- Queries based on pseudo-documents can be calculated based on the LSA space to assess similarity between pseudo-documents and actual documents represented in the space.

References

Berry, M. W., & Browne, M. (1999). *Understanding search engines: Mathematical modeling and text retrieval*. Philadelphia: Society for Industrial and Applied Mathematics.

Deerwester, S., Dumais, S. T., Furnas, G. W., Landauer, T. K., & Harshman, R. (1990). Indexing by latent semantic analysis. *Journal of the American Society for Information Science, 41*(6), 391.

Dumais, S. T. (1991). Improving the retrieval of information from external sources. *Behavior Research Methods, Instruments, & Computers, 23*(2), 229–236.

Dumais, S. T., Furnas, G. W., Landauer, T. K., & Deerwester, S. (1988). Using latent semantic analysis to improve information retrieval. In *Proceedings of CHI'88: Conference on Human Factors in Computing* (pp. 281–285). New York: ACM.

Griffiths, T. L., Steyvers, M., & Tenenbaum, J. B. (2007). Topics in semantic representation. *Psychological Review, 114*(2), 211–244.

Hu, M., & Liu, B. (2004, August). Mining and summarizing customer reviews. In *Proceedings of the Tenth ACM SIGKDD International Conference on Knowledge Discovery and Data Mining* (pp. 168–177). ACM.

Landauer, T. K., & Dumais, S. T. (1997). A solution to Plato's problem: The latent semantic analysis theory of acquisition, induction, and representation of knowledge. *Psychological Review, 104*(2), 211.

Landauer, T. K., Foltz, P. W., & Laham, D. (1998). An introduction to latent semantic analysis. *Discourse Processes, 25*(2–3), 259–284.

Martin, D. I., & Berry, M. W. (2007). Mathematical foundations behind latent semantic analysis. In *Handbook of latent semantic analysis*, 35–56.

Turney, P. D., & Pantel, P. (2010). From frequency to meaning: Vector space models of semantics. *Journal of Artificial Intelligence Research, 37*, 141–188.

Further Reading

For more about latent semantic analysis (LSA), see Landauer et al. (2007).

Chapter 7
Cluster Analysis: Modeling Groups in Text

Abstract This chapter explains the unsupervised learning method of grouping data known as cluster analysis. The chapter shows how hierarchical and k-means clustering can place text or documents into significant groups to increase the understanding of the data. Clustering is a valuable tool that helps us find naturally occurring similarities.

Keywords Cluster analysis · Hierarchical cluster analysis · k-means cluster analysis · k-means · Single linkage · Complete linkage · Centroid · Ward's method

7.1 Introduction

Cluster analysis is an unsupervised learning method, meaning that we do not know the true classification of a document in a document collection. Clustering methods apply algorithms based on distance or similarity measurements to group similar items together into clusters. As in latent semantic analysis (LSA), covered in Chap. 6, clustering reduces the dimensionality of the full document's representation by combining the terms or documents into groups. If we have one million documents, but are able to create a few thousand smaller groups or clusters, describing the documents becomes a much simpler task.

In cluster analysis, we try to find meaningful and interpretable groupings that naturally occur in the data. Clustering is used in information retrieval and indexing to aid in the categorization of unclassified data. In text analytics, clustering can be performed on either documents or terms to find similar groupings of either. An overview of the clustering process is shown in Fig. 7.1. In the context of text analysis, clustering can be used to find similarities in either document-document or term-term relationships in a term-document matrix (TDM) or document-term matrix (DTM). The input for the model is the TDM or DTM representation of the document collection. From there, the distance or similarity between terms-terms or documents-documents may be calculated and used as an input. Next, the analysis is conducted using one of two methods, hierarchical or

© Springer Nature Switzerland AG 2019
M. Anandarajan et al., *Practical Text Analytics*, Advances in Analytics and Data Science 2, https://doi.org/10.1007/978-3-319-95663-3_7

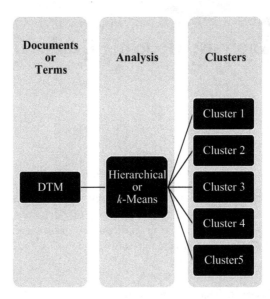

Fig. 7.1 Visualization of cluster analysis

k-means. The result of the analysis is a cluster solution in which terms or documents are assigned to clusters for evaluation by the analyst.

In this chapter, we will look specifically at these two approaches to clustering, k-means and hierarchical cluster analysis. In the examples in this chapter, we use the TDM created in Chap. 5, in which pet owners describe their pets. The TDM representation includes 12 terms and 10 documents.

7.2   Distance and Similarity

When we consider the distance between two items, we try to measure the difference or space between them. In this sense, we may want to consider how the two items differ across dimensions. If we consider an apple and banana, they differ in several dimensions: color, texture, taste, and size. Text-based cluster analysis relies on a similar notion, but the dimensions used to measure the distance, or difference, are either the terms or documents.

Clustering methods utilize distance or similarity measurements as the input for the clustering algorithms used to form the groups. We begin by plotting our terms in document space using our *tfidf*-weighted matrix. To better understand the concept of distance applied to text analytics, we can visualize terms in document space. Figure 7.2 displays five terms in document space for Documents 3 and 6. As shown, for these two documents, there appear to be some terms that are more similar than others. In particular, *fluffy* and *dog* are located close together, as are *favorite* and *hat*. In contrast, *brown* is not located near any of the plotted terms.

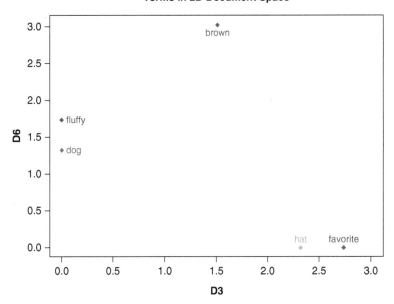

Fig. 7.2 Two-dimensional representation of terms in document space for Documents 3 and 6

Figure 7.3 expands this notion of similarity to three-dimensional space for Documents 1, 3, and 7. It is clear from this higher dimensional view that the groupings of terms change as more documents are considered. As more and more documentsv are considered in the distance calculation, the groupings will continue to change.

One common measure of the distance between two objects is Euclidean distance. The Euclidean distance, $d(x, y)$, between two points, x and y, in two dimensions can be calculated as

$$d(x,y) = \sqrt{(y_1 - x_1)^2 + (y_2 - x_2)^2}.$$

For example, as shown in Fig. 7.4, if we take the words *fluffy* and *brown* in two-dimensional document space for Documents 3 and 6, we can calculate the Euclidean distance between the two terms as

$$d(fluffy, brown) = \sqrt{(1.74 - 0.00)^2 + (3.03 - 1.51)^2} = 2.30.$$

The Euclidean distance between *fluffy* and *brown* in these two documents is equal to 2.30. We can calculate the distance between each of the terms in the document space by extending this two-dimensional distance metric to higher

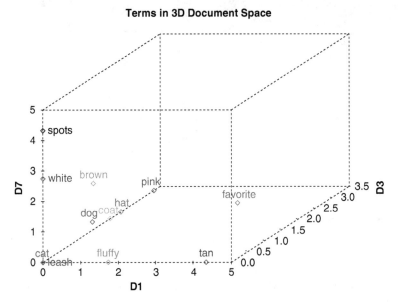

Fig. 7.3 Three-dimensional representation of terms in document space for Documents 1, 3, and 7

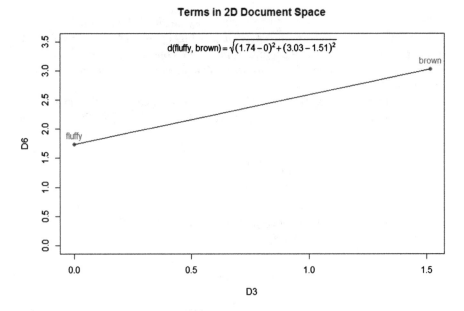

Fig. 7.4 *fluffy* and *brown* in document space for Documents 3 and 6

Table 7.1 Distance matrix of terms

	brown	cat	coat	dog	favorite	fluffy	hat	leash	pink	spots	tan	white
brown	0.0											
cat	4.7	0.0										
coat	5.0	7.1	0.0									
dog	4.1	5.1	5.5	0.0								
favorite	6.0	6.4	6.6	5.2	0.0							
fluffy	4.6	6.1	3.9	3.1	4.6	0.0						
hat	6.1	6.3	5.9	4.7	4.3	4.9	0.0					
leash	7.0	6.1	7.1	5.1	6.4	6.1	4.5	0.0				
pink	5.6	6.4	5.2	4.9	5.1	5.4	5.3	6.4	0.0			
spots	6.0	6.1	7.1	5.1	6.4	6.1	6.3	6.1	6.4	0.0		
tan	7.0	6.1	7.1	5.1	4.2	4.7	6.3	6.1	6.4	6.1	0.0	
white	6.0	6.4	5.7	3.6	6.7	4.6	5.6	6.4	5.1	4.2	6.4	0.0

Table 7.2 Distance matrix of documents

	D1	D2	D3	D4	D5	D6	D7	D8	D9	D10
D1	0.0									
D2	7.6	0.0								
D3	6.8	7.1	0.0							
D4	7.3	7.2	6.7	0.0						
D5	6.8	6.5	5.6	6.9	0.0					
D6	6.3	5.1	5.6	6.3	2.8	0.0				
D7	7.6	6.9	7.5	7.2	6.9	6.0	0.0			
D8	7.3	6.9	5.5	7.3	5.4	4.8	6.2	0.0		
D9	5.1	6.1	4.4	5.6	5.5	4.4	6.5	6.1	0.0	
D10	6.6	6.9	5.8	5.8	4.6	4.7	5.8	4.5	4.6	0.0

dimensions. The distances can be calculated in this way for all terms and documents to produce a distance matrix that captures the distance between words across all documents.

The distance matrices based on Euclidean distance for term-term distance and document-document distance are presented in Tables 7.1 and 7.2, respectively. In the analysis of the example, we will cluster the documents. Therefore, we will use the distance matrix in Table 7.2.

An alternative measure to calculating distance is calculating similarity. Recall from Chap. 6 that cosine similarity is used in latent semantic analysis (LSA) as a measure of closeness. Cosine similarity provides insights into the similarity between two documents or two terms based on the angle of the vectors that they form in LSA.

7.3   Hierarchical Cluster Analysis

Hierarchical cluster analysis (HCA) is a type of agglomerative clustering. Agglomeration methods iteratively join clusters together until there is one large document group with many smaller subgroups (Jain and Dubes 1988; Jain et al. 1999; Rousseeuw and Kaufman 1990). The result of hierarchical clustering analysis produces a visualization of the solution known as a dendrogram. A dendrogram is a tree-like diagram that depicts the hierarchical nature of the process. Cutting the dendrogram at different levels or heights leads to different clustering solutions.

The dendrogram displays the groupings created by the specific hierarchical clustering algorithm and the height at which the clusters are formed. By drawing horizontal lines on the dendrograms at varying heights, we make cuts to form the cluster groups. The number of clusters chosen for a specific cluster solution can be largely subjective in nature, although there are some tools to aid in decision-making, which we will take a closer look at in Sect. 7.4.

An example of a dendrogram appears in Fig. 7.5. The horizontal lines represent cuts, forming the different cluster configurations. The vertical lines represent the data that is being clustered, and the end of each of the lines is labeled as either a term or document. For instance, if we cluster documents, each vertical line represents a document. On the right-hand side, the cluster groupings are listed for each of the seven cuts, beginning with one-cluster solution, in which all documents are in the same cluster. HCA algorithms start at the bottom of the dendrogram, with all eight data points in their own clusters. As shown, larger and larger clusters are formed moving upward along the dendrogram. At the top of the dendrogram, one large cluster is created containing all eight data points. To see the impact of the dendrogram cuts, however, it is easier to start from the one-cluster solution and work down to the eight-cluster solution, in which each document is its own cluster. In Fig. 7.5, cuts are made at successively lower heights to form higher numbers of clusters.

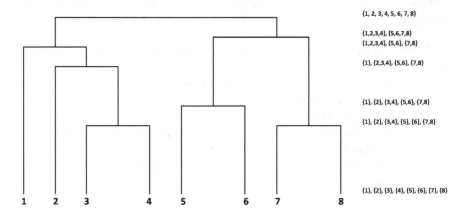

Fig. 7.5 Dendrogram example with cluster groupings

7.3.1   Hierarchical Cluster Analysis Algorithm

The general algorithm involves three main steps, with the fourth step being the repetition of previous steps in the process. In HCA, everything starts in its own cluster, and the singleton clusters are iteratively combined, based on the linkage method used. A general algorithm for agglomerative clustering, adapted from Xu and Wunsch (2005), is outlined in Fig. 7.6.

There are several linkage measures that can be used in Step 2 of the general algorithm to measure similarity or distance among clusters. These linkage measures can be separated into two groups: geometric methods and graph methods (Murtagh 1983). Geometric methods include centroid, median, and Ward's minimum variance method. Graph methods include single and complete-linkage methods. We begin by describing the graph methods.

7.3.2   Graph Methods

7.3.2.1   Single Linkage

If single linkage is used to combine clusters, the similarity between clusters is the maximum of the similarities (or the minimum distance) between all document pairs (Voorhees 1986). The dendrograms formed using single linkage tend to have long, stringy clusters.

In this method, to make clustering decisions, we seek to minimize the distance between clusters, $d(C_i, C_j) = \min\{d(x, y)|x \in C_i, y \in C_j\}$, where C_i and C_j are two

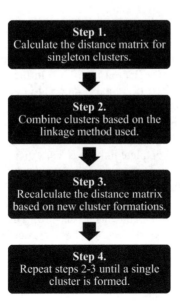

Fig. 7.6 HCA algorithm

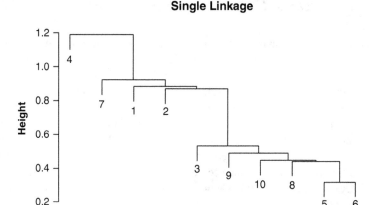

Fig. 7.7 Single linkage document—HCA example

different clusters, x is a member of C_i and y is a member of C_j. In Fig. 7.7, which depicts the dendrogram for a clustering solution based on single linkage, we see that, as expected, the first documents to be combined into a cluster are Documents 5 and 6. These have the smallest distance measurement in the distance matrix.

The advantage of single-linkage clustering is its efficient implementation. However, single linkage can sometimes lead to chaining based on transitive relationships, resulting in inappropriate groupings (Aggarwal and Zhai 2012; Jain et al. 1999; Nagy 1968).

7.3.2.2   Complete Linkage

In complete-linkage clustering, the maximum distance between a pair of documents is used in cluster formation (Zhao et al. 2005). In the complete-linkage method, distance is calculated as $d(C_i, C_j) = \max\{d(x, y) \mid x \in C_i, y \in C_j\}$, where C_i and C_j are two different clusters, x is a member of C_i and y is a member of C_j. This method creates more compact clusters than the single-linkage method and does not suffer from the same chaining issue (Jain et al. 1999; Jain et al. 2000). However, in comparison to single linkage, complete linkage is less efficient.

In the dendrogram in Fig. 7.8, we see more compact clusters. Again, the process starts by creating a cluster between the two closest documents, Documents 5 and 6. Next, instead of adding Document 8 to the cluster, as in single linkage, another cluster is formed, joining Documents 8 and 10. This process continues until Document 4 is finally joined to create the large cluster solution containing all ten documents.

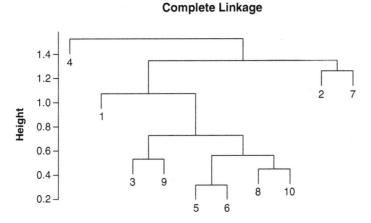

Fig. 7.8 Complete linkage document HCA example

7.3.3   Geometric Methods

7.3.3.1   Centroid

In clustering using centroid distance, the measurement of the dissimilarity or distance between clusters is the distance between the clusters' centers or centroids (Berry and Linoff 2011). Put simply, the similarity is calculated based on the middle of the clusters. Centroid-based methods are a type of average-link clustering. In this method, combination decisions are made, and distance is calculated as $d\left(C_{i \neq i}, C_{i \neq i}\right) = d\left(\mu_{i=i}, \mu_{i \neq i}\right)$, where μ_i is the mean or average of Cluster i and $\mu_i = \frac{1}{n}\sum x_k$. x_k represents a document in Cluster i.

As Fig. 7.9 illustrates, centroid-based methods can have issues with crossovers, which single linkage and complete linkage do not have. This crossover occurs when a cluster merges with another cluster, before the original cluster is even created (Jain and Dubes 1988). The overlaps shown on the dendrogram depict this crossover effect.

7.3.3.2   Ward's Minimum Variance Method

Ward's method or Ward's minimum variance method minimizes the within-cluster variance (Ward 1963). Cluster formation decisions are made based on calculations of squared errors, with the goal of minimizing the change in the sum of squared errors (SSE) caused by merging the clusters. There are many examples demonstrating that Ward's method performs better than alternative hierarchical clustering methods (Jain and Dubes 1988). In Ward's method, the SSE of Cluster i is calculated based on the distance from the cluster centroids. The cluster centroids are

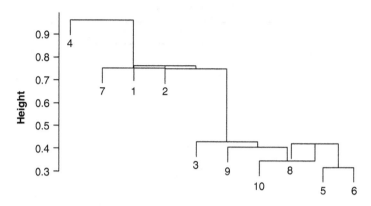

Fig. 7.9 Centroid linkage document—HCA example

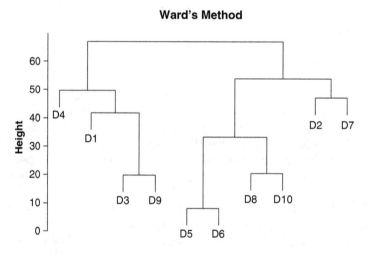

Fig. 7.10 Ward's method for linking documents—HCA example

calculated as in the centroid method. If there are m total clusters and n total docu-ments, the SSE of cluster i is computed as $SSE_i = \sum_{k=1}^{n}\sum_{i=1}^{m}\left[x_k - \mu_i\right]^2$, and the distance between clusters is found by calculating the change in the overall SSE of the cluster-ing solutions. For this reason, the squared Euclidean distance is used.

Figure 7.10 illustrates the resulting dendrogram from HCA using Ward's method. In this example, Ward's method provides a clear visualization of the clustering of the documents.

7.3.4   *Advantages and Disadvantages of HCA*

Hierarchical cluster analysis has many beneficial properties for unsupervised text categorization. Unlike other forms of clustering, such as *k*-means, which will be covered next, HCA does not require any predetermined number of clusters prior to performing the analysis. Additionally, HCA can be used as the only method of clustering, or it can be used as the basis for alternative forms of clustering. Another important advantage of HCA is that it produces a visualization, the dendrogram. In smaller samples, the dendrogram can be a clear representation of the analysis and can be easy to interpret, which is useful in presenting the results of the analysis.

Despite the many advantages of HCA, there are some known disadvantages too. In hierarchical clustering applied to text, the visualization can be difficult to interpret in large document collections. As the dimensionality of the analysis increases, the readability and usefulness of the dendrogram decrease. In large-scale analysis, the explanatory power of the dendrogram may not be as strong, without a reduction in dimensions or feature selection methods being used prior to analysis. Due to the high level of dimensionality of real-world text analysis, the use of the dendrogram by the analyst may not be as informative as in smaller-scale applications.

Another issue is that HCA can be particularly sensitive to outliers, whose presence may have a strong influence on the clustering solution (Manning et al. 2008). For this reason, it is sometimes beneficial to consider feature selection procedures and outlier detection prior to conducting HCA. While the fact that there is no need to specify the number of clusters at the outset is an advantage of hierarchical clustering, it can also lead to overly subjective results that are heavily dependent on the decisions of the analyst. For this reason, it is important to make informed decisions. We will consider some important decisions at the end of this chapter, in Sect. 7.4.

7.4   *k*-Means Clustering

k-means clustering (kMC) is a type of partitional clustering in which there is a single solution. Unlike in HCA, where the number of clusters is chosen after the analysis produces many possible cluster solutions, kMC requires that the number of clusters, *k*, be determined prior to the analysis. *k*-means clustering in text analytics is also a method for finding natural groupings based on a TDM or DTM, but there is no hierarchical structure or dendrogram-type visualization created. Instead, groupings are formed around pre-specified cluster seeds based on similarity or minimum distance. There are a few variations of kMC analysis, namely, *k*-medians, *k*-mediods, and *k*-modes clustering analysis. However, we will present the most popular, *k*-means clustering analysis. Next, we describe the general two-step algorithm used in kMC, which involves assigning data points to clusters and updating distance measures iteratively until a desired threshold is met.

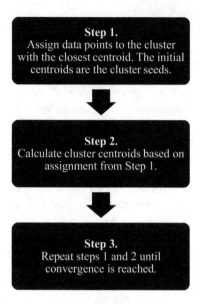

Fig. 7.11 kMC algorithm

7.4.1 kMC Algorithm

The *k*-means clustering method requires two inputs provided by the user: the number of clusters, *k*, and the initial seed. The initial seeds provide a starting point around which the clustering algorithm will build clusters. This initial seed can be data points or points chosen at random. In Sect. 7.4, we take a closer look at methods for choosing the initial seed. After the initial cluster seeds are chosen, the process follows two steps until all assignments are made, and based on some convergence criteria, the model cannot be improved.

Typically, we try to minimize the total sum of squared errors (SSE), as in Ward's method of HCA. The within-cluster SSE is calculated as the sum of the squared distance from each data point to its assigned cluster centroid. The sum of the within-cluster SSE values for all *k* clusters is the total SSE. When the total SSE cannot be improved by reassigning data points, the algorithm stops. The kMC algorithm is depicted in Fig. 7.11.

7.4.2 The kMC Process

To explain how kMC works, we perform a small-scale kMC analysis to cluster terms in the TDM. Specifically, we will apply the kMC algorithm to eight terms across two documents, Documents 3 and 6.[1] Their *tfidf*-weighted values are shown in Table 7.3.

---

[1] The omitted terms (*leash*, *spots*, *tan*, and *white*) have the same value as *cat* (0,0) and, therefore,

Table 7.3 *Tfidf* term values
for Documents 3 and 6

Term	D3	D6
brown	1.5	3.0
cat	0.0	0.0
coat	2.0	2.0
dog	0.0	1.3
favorite	2.7	0.0
fluffy	0.0	1.7
hat	2.3	0.0
pink	3.3	0.0

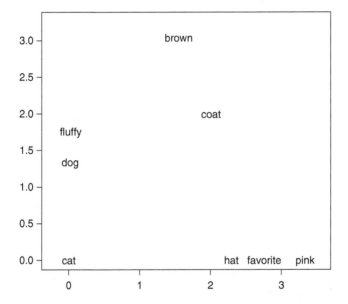

Fig. 7.12 *k*-Means process example plot

Figure 7.12 shows the plotted *tfidf* term values for the two documents. Based on the plotted values, we have reason to believe that there are three naturally occurring groupings consisting of {*brown, coat*}, {*fluffy, dog, cat*}, and {*hat, favorite, pink*}. For this reason, we will choose *k* = 3. In larger *k*-means clustering applications, especially with high dimensionality, this visual decision method cannot be used. In Sect. 7.4.1, we will introduce some methods for choosing *k*.

Prior to kMC, we need to choose (1) *k*, the number of clusters, (2) the initial cluster seeds, and (3) the convergence criteria. The three initial seeds are chosen to be three randomly selected data points: *cat, coat*, and *hat*. Figure 7.13 designates these cluster seeds with a purple "X" as the plot point. These seeds serve as our

---

would be in the same cluster. To make the example visually clearer, they were removed.

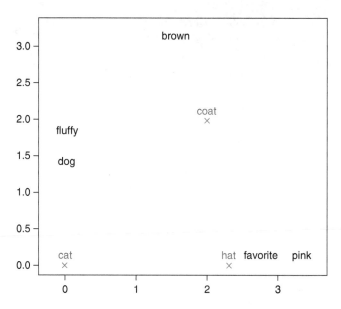

Fig. 7.13 k-Means initial cluster seed designation of *cat*, *coat*, and *hat*

Table 7.4 Squared distance from cluster seeds

	Centroid 1: cat	Centroid 2: coat	Centroid 3: hat
brown	11.5	**1.3**	9.8
dog	**1.7**	4.5	7.1
favorite	7.5	4.5	**0.2**
fluffy	**3.0**	4.1	8.4
pink	11.0	5.7	**1.0**

initial cluster centroids, and all initial assignments are based on the closeness of the remaining data points to these cluster seeds.

Once the initial cluster seeds are chosen, the distances between the seeds and the remaining data points are calculated. Table 7.4 displays the squared distance between the unassigned data points shown in black in Fig. 7.13 and the cluster centroids plotted in purple. The assignment is chosen by evaluating the minimum distance for each of the rows in the table. For each term i, assignment$_i$ = argmin$_j\{d$ (term$_i$, centroid$_j)^2\}$. The minimum distance for each of the terms is shown in bold in the table.

Our initial cluster assignments, which are based on the minimum squared distance values, are depicted in Fig. 7.14. The terms in Cluster 1 are red, Cluster 2 are green, and Cluster 3 are blue. The within-cluster SSE is calculated by adding the squared distance values for each cluster, resulting in within-cluster SSE values of 4.7, 1.3, and 1.2, respectively. The total within-cluster SSE, therefore, is $4.7 + 1.3 + 1.2 = 7.2$.

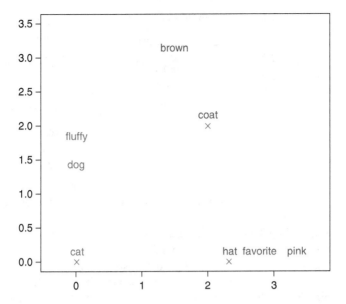

Fig. 7.14 Cluster assignments based on cluster seeds for a three-cluster solution

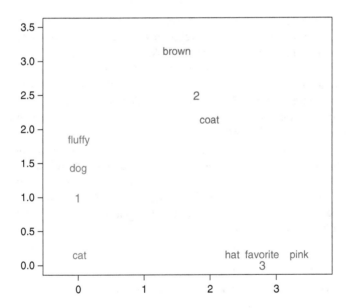

Fig. 7.15 First iteration cluster assignment and calculated centroids

The next step, Step 2, is to recalculate the centroids of the clusters. Since the centroid of a cluster is the arithmetic mean of the cluster, we can easily calculate the new cluster centroids as (0.0, 1.0), (1.8, 2.5), and (2.8, 0.0), for Clusters 1, 2, and 3,

Table 7.5 Squared distance from terms to cluster centroids

Term	Cluster 1	Cluster 2	Cluster 3
brown	6.4	**0.4**	10.8
cat	**1.0**	9.5	7.8
coat	5.0	**0.3**	4.6
dog	**0.1**	4.6	9.6
favorite	8.5	7.1	**0.0**
fluffy	**0.5**	3.8	10.9
hat	6.4	6.5	**0.2**
pink	12.0	8.6	**0.3**

respectively. These new centroid values are plotted in Fig. 7.15 and are labeled by their cluster number.

Next, based on Step 3, we repeat the process, beginning with Step 1 to determine if we can improve the assignments from our first iteration. For each of the terms, we recalculate the square distances to each of the cluster centroids. Table 7.5 presents the results. The minimum squared distance for each term is highlighted in bold.

The within-cluster SSE values for the three clusters are 1.5, 0.6, and 0.5, for a total SSE of 2.6. Comparing our new total SSE to our previous total SSE, we see that the total SSE has been reduced from 7.2 to 2.6. Based on Table 7.5, the assignments have remained unchanged. Thus, we can conclude that this is the best possible clustering solution based on the chosen k value and initial seeds. Since the solution remains unchanged, we stop the clustering algorithm iteration and choose this as our clustering solution. Therefore, our kMC analysis solution in this example is {*fluffy, dog, cat*}, {*brown, coat*}, and {*hat, favorite, pink*}. If this were not the case, and we could improve the solution, we would make the necessary assignments and continue iterations of the algorithm until the solution could not be improved any further based on the total within-cluster SSE.

Next, we consider the kMC cluster solution for the full document collection using the *tfidf*-weighted TDM created in Chap. 5. As in the small-scale example, we choose $k = 3$. Applying the kMC algorithm, Cluster 1 contains Documents 2 and 7; Cluster 2 contains Documents 1, 3, 4, and 9; and Cluster 3 contains Documents 5, 6, 8, and 10. The within-cluster SSE for the three clusters are 23.6, 55.6, and 30.9. The total within-cluster SSE is 110.

7.4.3 Advantages and Disadvantages of kMC

kMC analysis has several advantages in unsupervised text categorization. kMC performs fast calculations, and the algorithm typically converges quickly (Aggarwal and Zhai 2012). When there are clear clusters in the data, the performance is particularly good. Many of the shortcomings of k-means are related to the choices made prior to performing the analysis. The kMC method requires that the number of clusters be chosen before conducting the cluster analysis. In the next section, we cover some ways to choose the number of clusters. Additionally, different initial seed choices lead to very different clustering solutions. It is widely accepted that based on the choice of the initial seed value, the kMC algorithm can converge after finding local optima instead of global optima. Some researchers suggest running the k-means algorithm many times, using different initial seeds each time to evaluate many solutions, choosing the best solution of these runs (Steinley 2006). Depending on the size of the data, this approach may be one method to search for the best possible solution based on the minimized SSE. However, for larger datasets, this method may be infeasible.

7.5 Cluster Analysis: Model Fit and Decision-Making

7.5.1 Choosing the Number of Clusters

The number of clusters is an important decision in both cluster analysis methods. We present some subjective methods and some methods based on graphing fit measures to choose either the minimum or maximum fit value. Since both HCA and kMC are sensitive to the decisions of the analyst, it is important to consider many alternatives when choosing the final clustering solution. There is no single objective number of optimal clusters, so a thorough evaluation of alternative solutions is key to finding the best model for the specific application.

7.5.1.1 Subjective Methods

In HCA, the dendrogram helps us visualize the solutions for varying numbers of clusters, k. First, we look at the cluster solution using Ward's method and consider different cluster analysis solutions with varying numbers of clusters, k. Using the dendrograms for each of the potential number of clusters helps us visualize the solutions for different numbers of clusters. Figure 7.16 displays clustering solutions in which k varies between 2 and 7. In this case, there is still subjectivity in the cluster decision. One person may believe that the solution with six clusters appears to fit best, while another person may believe that a two-cluster solution is best. For this reason, this method is more subjective than objective.

There are other subjective methods that can be utilized, such as using expert opinion, domain knowledge, or experience to choose the number. In k-means clustering, it can be beneficial to use the best solution from hierarchical clustering to determine the number of clusters as the input for k-means. For large datasets, the use of the dendrogram may not be sufficient to make the decision. However, there are several measures that can be utilized to aid in the decision.

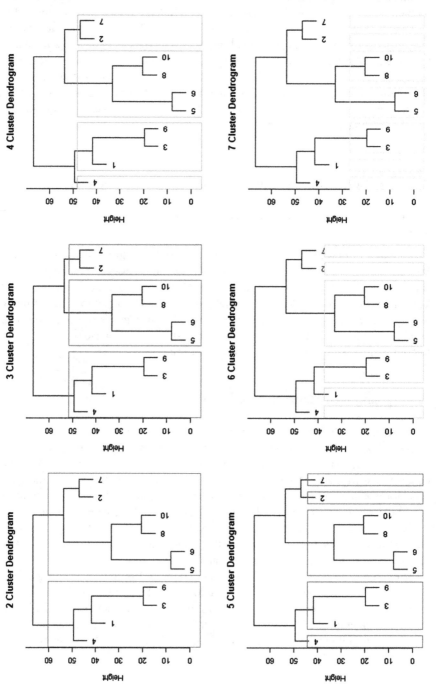

Fig. 7.16 Ward's method hierarchical cluster analysis solutions for $k = 2, 3, 4, 5, 6,$ and 7

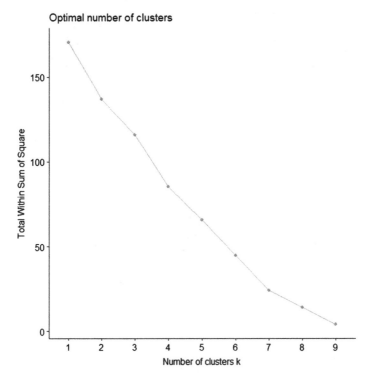

Fig. 7.17   Scree plot total within cluster SSE for k Values 1–9

7.5.1.2   Graphing Methods

Scree Plot

Using the kMC solution described in Sect. 7.3.2, we run the clustering algorithm
and vary the number of clusters, k, and evaluate the total within-cluster SSE. Plotting
k against the total within-cluster SSE, we create a plot that is similar to a scree plot.
A scree plot depicts the variance explained on the y-axis for different solutions
when varying the number of clusters, k, and is used widely in data mining analysis.
As in the case of a scree plot, we aim to find an elbow, to find a balance between the
explanatory power and the complexity of the solution. Figure 7.17 displays this plot
for the k-means solution. As evidenced by this plot, sometimes there is no clear
elbow. Even if there is an elbow, there is still subjectivity in the interpretation and
resulting decision based on this type of plot.

Silhouette Plot

The average silhouette method can be used with any clustering approach to help
choose the number of clusters (Rousseeuw and Kaufman 1990). The silhouette
coefficient, s_i, for data point i is calculated by

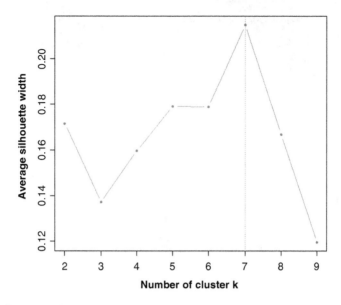

Fig. 7.18 Silhouette plot for Ward's method hierarchical clustering analysis

$$s_i = \frac{b_i - a_i}{\max\left(a_i, b_i\right)},$$

where a_i is the average distance from data point i to all other data points in its own cluster and b_i is determined by calculating the distance to all other data points in all other clusters and finding the minimum average distance from the data point to another cluster. Due to the need to compute a_i and b_i for each data point, it is clear that the calculation of the cluster silhouette coefficient can be computationally expensive for larger datasets.

From there, the average cluster silhouette is found by averaging the data point silhouette coefficients for each cluster. The overall measure of the average silhouette for the cluster solution is calculated as the average of the silhouette values for all points in the data. The average overall silhouette model values for k values between 2 and 9 for the Ward's method HCA is displayed in Fig. 7.18. Based on the plot, we would conclude that the best clustering solution would be the $k = 7$ clustering solution. The seven-cluster solution is {1}, {4}, {2}, {7}, {3,9}, {5,6}, and {8,10}.

7.5.2   Naming/Describing Clusters

In addition to finding naturally occurring groupings in the data, the main objective in the application of cluster analysis is to describe the clusters. It is important to consider what the members of clusters have in common and what differentiates them. The most useful clustering analysis solution should be interpretable.

We consider the *k*-means clustering solution with three clusters, which is identical to the three-cluster solution using HCA with Ward's method and seeks to describe the clusters by first looking at their centroids. The three-cluster *k*-means solution is {2,7}, {1,3,4,9}, and {5,6,8,10}.

Since the *tfidf* weightings are used as the inputs for the clustering solution, the importance of the various terms in the documents is what creates the similarities and dissimilarities in the documents. Recall that in the kMC example clustering terms, the centroids were in two dimensions, Documents 5 and 6. If the terms in the TDM were to be clustered, the centroids would be in ten dimensions, since the total number of documents in the collection is ten. In the larger example, the centroids are the number of dimensions of the terms used to create the clusters. In this case, the centroids would be in 12 dimensions, since there are 12 terms.

If we characterize our clusters by their most prominent terms, we can describe Cluster 1 using the words *brown*, *cat*, and *spots*, we can describe Cluster 2 using *favorite* and *hat*, and we can describe Cluster 3 using *coat* and *fluffy*. In real-world cases with clearly defined distinctions or categories in the data, describing the centroids in this way can provide more concrete labels or distinctions to the clusters. This ability is particularly helpful in presenting and reporting the analysis clearly.

7.5.3   Evaluating Model Fit

In validating our model to assess model goodness and fit, we are concerned with three types of validity: internal, external, and relative. Internal validity is based on measures derived from the data and the model created. External validity is based on outside information, such as expert opinion or classification labels. Finally, relative validity compares different solutions by changing the values of the parameters (Zaki et al. 2014). Due to the unsupervised nature of clustering in this chapter, we will focus on assessing the internal validity of the model. However, if we know the actual cluster membership, we can evaluate the predicted and actual cluster membership. The silhouette coefficient described in Sect. 7.4.1 is an example of a relative validity measure because choosing *k*, the number of clusters, is a parameter choice. In Chap. 9, which introduces supervised classification analysis methods, we cover external validity in depth.

One measure of internal validity used to assess the fit of a clustering solution is the Dunn index (Halkidi et al. 2002). The Dunn index is the ratio of the minimum inter-cluster data point distance to the maximum intra-cluster point distance or

$$\text{Dunn Index} = \frac{\min\{d_{\text{inter}}\}}{\max\{d_{\text{intra}}\}},$$

where min $\{d_{\text{inter}}\}$, the minimum inter-cluster distance, is the minimum distance between any two points in different clusters and max $\{d_{\text{intra}}\}$, the maximum intra-cluster distance, is the maximum distance between any two points in the same cluster. These distance measures are found by evaluating the distance measure used to create the model. Better fitting clustering models will have higher Dunn index values.

7.5.4 Choosing the Cluster Analysis Model

The choice between k-means and hierarchical clustering depends largely on the purpose of the application. When choosing a model type to use, consider the advantages and disadvantages of each. Since clustering is an unsupervised, undirected form of analysis, the choice of k in the kMC method can be difficult. For this reason, a combination of the two methods may be helpful in finding the naturally occurring groupings that exist in the dataset. Expertise or access to experts in the field who have extensive knowledge about the data can help inform the analysis and therefore the decision about the type of analysis.

A good understanding of the data will help determine the analysis method to use in any text analytic application. Clustering will always produce a solution, and in some cases, clusters will be created where they do not naturally occur in the data (Manning et al. 2008). If the data are too noisy, there may not be naturally occurring groupings in the TDM or DTM. Preprocessing and feature selection can reduce the dimensionality and help the analyst decide if cluster analysis is the right fit. In addition, the goal of the analysis should help in this decision. If the goal is to categorize documents in a document collection without knowing what the true categories are, cluster analysis is a good analysis method. If the true categories of the documents are known and the goal is to make predictions, a better approach is to use a supervised method, such as classification analysis, which will be introduced in Chap. 9.

Key Takeaways
- Cluster analysis is an unsupervised analysis method used to form groups, or clusters, based on similarity or distance.
- Two popular clustering methods are hierarchical cluster analysis (HCA) and k-means clustering (kMC).
- HCA produces a visualization of the clusters, known as a dendrogram, and can be completed using several methods, including single-linkage, complete-linkage, centroid, and Ward's method.
- kMC requires cluster seeds, the number of clusters, and a convergence criterion as input to the analysis.
- Two types of methods to choose the number of clusters are subjective and graphing.

References

Aggarwal, C. C., & Zhai, C. X. (2012). *Mining text data.* New York: Springer Verlag.

Berry, M. J., & Linoff, G. S. (2011). *Data mining techniques. For marketing, sales, and customer relationship management.* Chichester: Wiley-Interscience.

Halkidi, M., Batistakis, Y., & Vazirgiannis, M. (2002). Cluster validity methods: Part II. *ACM SIGMOD Record, 31*(2), 40–45.

Jain, A. K., & Dubes, R. C. (1988). *Algorithms for clustering data.* Upper Saddle River: Prentice-Hall.

Jain, A. K., Murty, M. N., & Flynn, P. J. (1999). Data clustering: A review. *ACM Computing Surveys (CSUR), 31*(3), 264–323.

Jain, A. K., Duin, R. P. W., & Mao, J. (2000). Statistical pattern recognition: A review. *IEEE Transactions on Pattern Analysis and Machine Intelligence, 22*(1), 4–37.

Manning, C., Raghavan, P., & Schütze, H. (2008). *Introduction to Information Retrieval.* *Cambridge:* Cambridge University Press. doi: 10.1017/CBO9780511809071.

Murtagh, F. (1983). A survey of recent advances in hierarchical clustering algorithms. *The Computer Journal, 26*(4), 354–359.

Nagy, G. (1968). State of the art in pattern recognition. *Proceedings of the IEEE, 56*(5), 836–863.

Rousseeuw, P. J., & Kaufman, L. (1990). *Finding groups in data.* Hoboken: Wiley Online Library.

Steinley, D. (2006). K-means clustering: A half-century synthesis. *British Journal of Mathematical and Statistical Psychology, 59*(1), 1–34.

Voorhees, E. M. (1986). Implementing agglomerative hierarchic clustering algorithms for use in document retrieval. *Information Processing & Management, 22*(6), 465–476.

Ward, J. H., Jr. (1963). Hierarchical grouping to optimize an objective function. *Journal of the American Statistical Association, 58*(301), 236–244.

Xu, R., & Wunsch, D. (2005). Survey of clustering algorithms. *IEEE Transactions on neural networks, 16*(3), 645–678.

Zaki, M. J., Meira, W., Jr., & Meira, W. (2014). *Data mining and analysis: Fundamental concepts and algorithms.* Cambridge: Cambridge University Press.

Zhao, Y., Karypis, G., & Fayyad, U. (2005). Hierarchical clustering algorithms for document datasets. *Data Mining and Knowledge Discovery, 10*(2), 141–168. https://doi.org/10.1007/s10618-005-0361-3.

Further Reading

For more about clustering, see Berkhin (2006), Jain and Dubes (1988) and Jain et al. (1999).

Chapter 8
Probabilistic Topic Models

Abstract In this chapter, the reader is introduced to an unsupervised, probabilistic analysis model known as topic models. In topic models, the full TDM (or DTM) is broken down into two major components: the topic distribution over terms and the document distribution over topics. The topic models introduced in this chapter include latent Dirichlet allocation, dynamic topic models, correlated topic models, supervised latent Dirichlet allocation, and structural topic models. Finally, decision-making and topic model validation are presented.

Keywords Topic models · Probabilistic topic models · Latent Dirichlet allocation · Dynamic topic models · Correlated topic models · Structural topic models · Supervised latent Dirichlet allocation

8.1 Introduction

Topic models, also referred to as probabilistic topic models, are unsupervised methods to automatically infer topical information from text (Roberts et al. 2014). In topic models, topics are represented as a probability distribution over terms (Yi and Allan 2009). Topic models can either be single-membership models, in which documents belong to a single topic, or mixed-membership models, in which documents are a mixture of multiple topics (Roberts et al. 2014). In this chapter, we will focus on mixed-membership models. In these models, the number of topics, k, is a fixed number that is chosen prior to building the model.

Latent semantic analysis (LSA), which is covered in Chap. 6, and topic models are both dimension reduction methods and use the document-term matrix (DTM) or term-document matrix (TDM) as the input for the analysis. While LSA discovers hidden semantic content, topic models reveal thematic structure. LSA aims to uncover hidden meaning in text, while topic models focus on the underlying subjects or themes that are present in the documents.

The most common type of topic model was created as an extension of the probabilistic latent semantic indexing (pLSI) model proposed by Hofmann (1999), which is a probabilistic LSA model. Figure 8.1 shows how the specific dimension

© Springer Nature Switzerland AG 2019
M. Anandarajan et al., *Practical Text Analytics*, Advances in Analytics and Data Science 2, https://doi.org/10.1007/978-3-319-95663-3_8

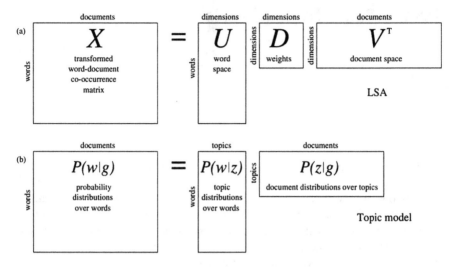

Fig. 8.1 LSA and topic models (Griffiths et al. 2007, p. 216)

reduction in LSA compares to that of topic models (Griffiths et al. 2007). As detailed in Chap. 6, LSA uses singular value decomposition (SVD) to break down the full TDM or DTM into three smaller component matrices. From there, the number of singular vectors can be reduced to create a smaller dimensional representation of the original. In topic modeling, the full TDM or DTM is broken down into two major components: the topic distribution over terms and the document distribution over topics. The first component tells us the importance of the terms in topics, and using that importance information, the second component tells us the importance of topics in the documents.

While there are many similarities between LSA and topic models, they also differ in many ways. Unlike LSA, topic models are generative probabilistic models. Based on the assigned probability, we can understand topics through their most likely terms. We are also able to better understand documents based on their most likely topics in their topic distribution. Unlike the latent factors in LSA, each topic is clearly identified and explainable.

The topic modeling examples in this chapter use text data based on 56 documents in which people describe the physical appearance of their pet dogs. There are four different breeds of dogs described in the documents: Bichon Frise, Dachshund, Great Dane, and Golden Retriever. The dogs vary in terms of height, weight, size, color, and fur. Each of the breeds has distinguishing features that are characteristic of that breed. For instance, Bichon Frises are small, fluffy dogs that are predominantly white and usually have black noses. On the other hand, a Great Dane is a very large dog with a short, straight coat that can be one of the several colors or a mix of colors. In the document sample, each dog type is described in 14 of the 56 documents.

Preprocessing and parsing are applied to the 56 text documents, including the removal of stop words and stemming. Since some of the pet owners describe their dogs in terms of weight, we do not remove numbers from our corpus. The *tfidf*-weighted DTM is used, and the resulting vocabulary contains 96 terms. The first topic model that we will present is the latent Dirichlet allocation (LDA). Many of the alternative topic models use LDA as the basis of their model.

8.2 Latent Dirichlet Allocation (LDA)

The latent Dirichlet allocation (LDA) model is a generative probabilistic model introduced in Blei et al. (2002, 2003). LDA assumes a bag-of-words (BOW) model representation, meaning that term ordering in a document is not considered when building the topic model (Blei et al. 2003). Additionally, LDA assumes that documents are exchangeable, meaning that there is no meaningful sequential ordering of the documents in the collection (Blei et al. 2010). Another assumption of LDA is the independence of topics. Figure 8.2 provides an illustration of the LDA.

In the LDA model, K is the total number of topics, D is the total number of documents, and N is the total number of words in a document, where $W_{d, n}$ is an observed word. Additionally, α is the Dirichlet parameter, and η is the topic hyperparameter. Each topic is a distribution over terms, with topic assignments $Z_{d, n}$. Each document is a mixture of topics, with topic proportions θ_d, and each term is drawn from one of the topics, with topic assignments β_k. Due to the intractability of computing the posterior distribution of the topics in a document, approximation methods are used, including mean field variational methods, expectation propagation, collapsed Gibbs sampling, and collapsed variational inference.

Using the data in the example, we build a model with four topics. We choose four topics as our starting point because we know that there are four dog breeds represented in the document collection. The top ten terms in each of the four topics based on the expected topic assignment are depicted in Fig. 8.3. Based on the figure, Topic 2 can be described using the word *white*, and we would expect to see the documents describing Bichon Frises assigned to this topic. Topic 1 can be described by the terms *long*, *tail*, and *short*; Topic 3 can be described by the terms *weigh*, *pound*, and *coat*; and Topic 4 can be described by *coat* and *ear*. One of the strengths of topic models can be seen in Topic 3. The terms *weigh* and *pound* are

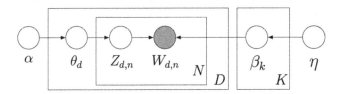

Fig. 8.2 Plate representation of the random variables in the LDA model (Blei 2012, p. 23)

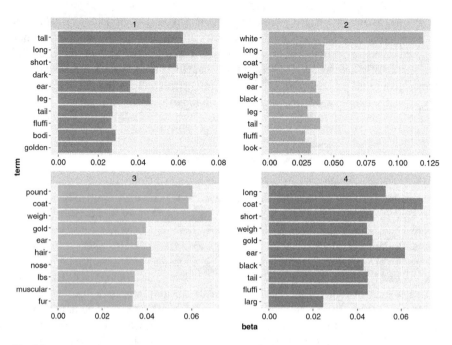

Fig. 8.3 Top ten terms per topic, four-topic model

synonyms, and they are most likely terms in the same topic. Unlike other methods, such as LSA, topic models are able to handle synonymous terms very well.

8.3 Correlated Topic Model (CTM)

The correlated topic model (CTM) is a hierarchical model that explicitly models the correlation of latent topics, allowing for a deeper understanding of relationships among topics (Blei and Lafferty 2007). The CTM extends the LDA model by relaxing the independence assumption of LDA. As in the LDA model, CTM is a mixture model and documents belong to a mixture of topics. CTM uses the same methodological approach as LDA, but it creates a more flexible modeling approach than LDA by replacing the Dirichlet distribution with a logistic normal distribution and explicitly incorporating a covariance structure among topics (Blei and Lafferty 2007). While this method creates a more computationally expensive topic modeling approach, it allows for more realistic modeling by allowing topics to be correlated. Additionally, Blei and Lafferty (2007) show that the CTM model outperforms LDA (Fig. 8.4).

As in the LDA model, K is the total number of topics, D is the total number of documents, and N is the total number of words in a document, where $W_{d,n}$ is an observed word. In the CTM model, η_d is the topic hyperparameter with mean μ and

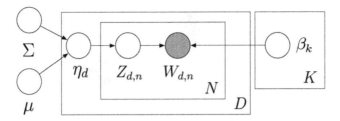

Fig. 8.4 Plate representation of the random variables in the CTM model (Blei et al. 2007, p. 21)

Top Topics

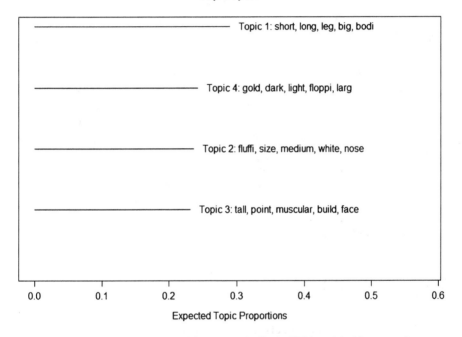

Fig. 8.5 Expected topic proportions of four categories in the CTM model with no covariates

covariance matrix $\sum$. Again, each topic is a distribution over terms, with topic assignments $Z_{d,n}$. Each document is a mixture of topics, with topic proportions θ_d, and each term is drawn from one of the topics, with topic assignments β_k. A fast variational inference algorithm is used to estimate the posterior distribution of the topics, because, as in LDA, the calculation is intractable. However, in practice, the computation is inefficient, particularly in comparison to LDA.

Using the data from the example, we build a CTM model with $k = 4$ topics. The top terms and expected topic proportions of this model are presented in Fig. 8.5. When considering the topic proportions, since we know that the four dogs are equally represented in the document collection, we would expect the topics to have the same expected proportions if the topics are dog specific. Based on the figure,

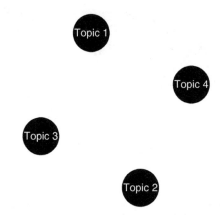

Fig. 8.6 CTM topic correlation plot

they do not appear to be topics based on the dog breeds. However, it does appear that Topic 1 could be about Dachshunds, Topic 2 about Golden Retrievers, Topic 3 about Bichon Frises, and Topic 4 about Great Danes. Topic 1 is the most prevalent expected topic, which contains *short*, *long*, and *leg* as the topic words. To try to explain the difference in topic proportions, we can look at a plot of the correlations among topics.

The CTM model has the advantage over the LDA model in that it models the correlations among topics. To investigate possible correlations, we can evaluate the correlations among topics and create an adjacency plot. Based on the adjacency plot in Fig. 8.6, in which no edges or straight lines connect the topic nodes, the four topics in the CTM model are not correlated.

8.4   Dynamic Topic Model (DT)

The dynamic topic model models topics in a sequentially ordered document collection to incorporate the evolution of topics over time by relaxing the exchangeability assumption of the LDA model (Blei and Lafferty 2006). The process involves splitting the data into smaller, time-dependent groups, such as by month or year. Dynamic topic models are built as an extension of the LDA model and thus do not model correlations among topics. The model uses the logistic normal distribution with mean α for each time period t (Fig. 8.7).

As in the LDA model, K is the total number of topics, D is the total number of documents, and N is the total number of words in a document. $W_{d,n}$ is an observed word. Additionally, α_t is the mean Dirichlet parameter α at time t. Each topic is a distribution over terms, with topic assignments $Z_{t,d,n}$. Each document is a mixture of topics, with topic proportions $\theta_{t,d}$, and each term is drawn from one of the topics, with topic assignments $\beta_{t,k}$ at time t. The model can use variational Kalman filtering or a variational wavelet regression to estimate the parameters of the DT model.

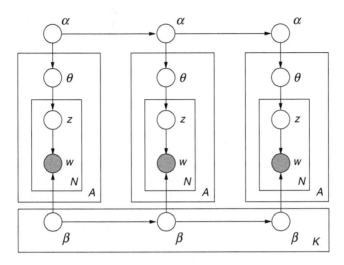

Fig. 8.7 Plate diagram of DT model (Blei and Lafferty 2006, p. 2)

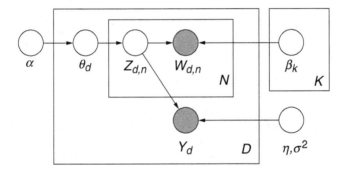

Fig. 8.8 Plate representation of the sLDA model (McAuliffe and Blei 2008, p. 3)

8.5   Supervised Topic Model (sLDA)

McAuliffe and Blei (2008) introduced the supervised latent Dirichlet allocation (sLDA), which is an extension of the LDA model with the use of labeled documents, as in the classification analysis covered in Chap. 9. The sLDA model has a class variable associated with each document, which serves as the response variable in the model (Fig. 8.8).

As in the LDA model, in the sLDA model, K is the total number of topics, D is the total number of documents, and N is the total number of words in a document, where $W_{d,n}$ is an observed word. Additionally, α is the Dirichlet parameter, η and σ^2 are response parameters, and y is the response variable. Each topic is a distribution over terms, with topic assignments $Z_{d,n}$. Each document is a mixture of topics, with topic proportions θ_d, and each term is drawn from one of the topics, with topic assignments β_k. Rather than treat the parameters as random variables, the model treats them as unknown constants. As in the LDA model, a variational expectation-maximization (VEM) procedure is used for the model estimation.

8.6   Structural Topic Model (STM)

The structural topic model (STM) combines three common topic models to create a semiautomated approach to modeling topics, which can also incorporate covariates and metadata in the analysis of text (Roberts et al. 2014). Additionally, unlike the LDA model, topics in STM can be correlated. This model is particularly useful in the topical analysis of open-ended textual data, such as survey data.

Since STM allows for the addition of covariates, additional information from the data can be used in the model. Furthermore, effect estimation can be performed to investigate and compare selected covariates and topics. In particular, STM has the ability to account for topical content and prevalence, allowing us to compare groupings in the data. For instance, to consider content, we could compare the specific words that are used to describe the different types of dogs. We could also explore the topic's prevalence, or how often a topic occurs, for the different breeds.

STM is a mixture model, where each document can belong to a mixture of the designated k topics. Topic proportions, θ_d, can be correlated, and the topical prevalence can be impacted by covariates, X, through a regression model $\theta_d \sim \text{LogisticNormal}(X\gamma, \Sigma)$. This capability allows each document to have its own prior distribution over topics, rather than sharing a global mean. For each word, w, the topic, $z_{d,n}$, is drawn from a response-specific distribution. Conditioned on the topic, a word is chosen from a multinomial distribution over words with parameters, $\beta_{zd,n}$. The topical content covariate, U, allows word use within a topic to vary by content (Fig. 8.9).

We build an STM model with four topics and include the dog breed as a content variable. The top ten words in each topic in the STM model are shown in Fig. 8.10.

By incorporating the dog breed as a covariate, we can consider how the breeds vary for each of the four topics. Figure 8.11 shows the expected proportion of topics for each of the topics and dog breeds.

8.7   Decision Making in Topic Models

8.7.1   Assessing Model Fit and Number of Topics

Although there is no single, uniform measure for choosing the number of topics in building a topic model, several methods have been proposed to help the analyst decide on the number of topics, k. Two methods aim to minimize the metrics to determine the optimal number of topics. Cao Juan et al. (2009) uses minimum

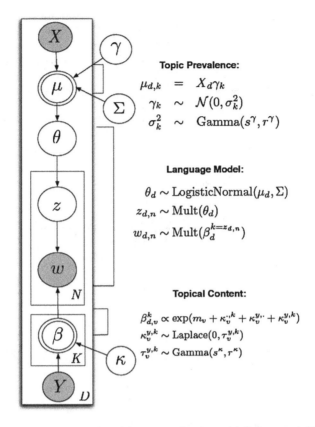

Fig. 8.9 Plate diagram representation of the structural topic model (Roberts et al. 2013, p. 2)

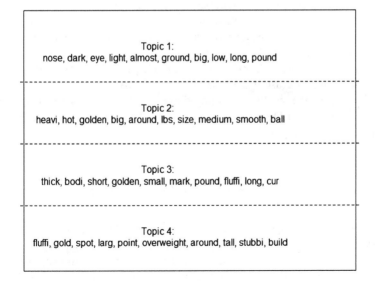

Fig. 8.10 Top ten terms in topics for STM model

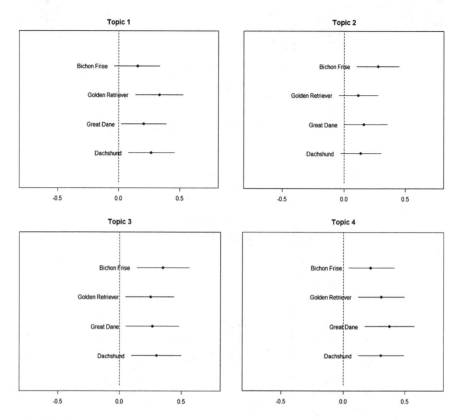

Fig. 8.11 Dog type content across topics

density measures to choose the number of topics. Arun et al. (2010) utilize a measure of divergence, where minimal divergence within a topic is preferred. Both methods use measures of distance to make decisions regarding k. On the other hand, Deveaud et al. (2014) utilize a measure maximizing the divergence across topics, and Griffiths and Steyvers (2004) maximize the log-likelihood of the data over different values of k. We use these four measures across 2–30 topics in building the LDA models, and the results are displayed in Fig. 8.12. Based on the figure, including five topics in an LDA model appears to be a good tradeoff between the four measures and is highlighted in red.

8.7.2 Model Validation and Topic Identification

Topic models can be supervised or unsupervised and, thus, can rely on either internal or external validity measures, depending on the type of data being used. Model validity and interpretability should go hand-in-hand in topic model analysis, and therefore, we will consider them together. We will focus on internal validity

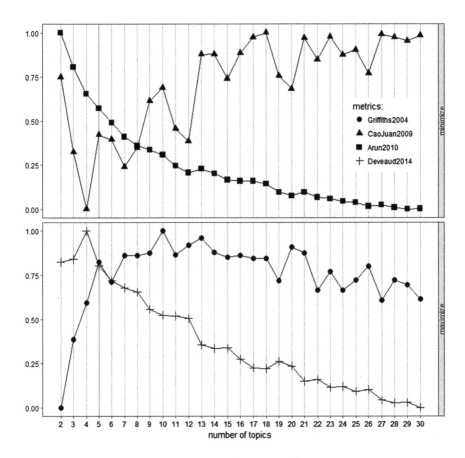

Fig. 8.12 Four measures across a number of topics, k, for 2–30 LDA topics

measures, since we performed an unsupervised LDA analysis. Mimno et al. (2011) suggest using topic size or the frequency of terms assigned to the topic as a good indicator of topic quality. Figure 8.13 displays the term frequency for the four-topic LDA solution and the five most probable terms in those topics.

Topics 4, 3, and 1, respectively, are the topics with the highest number of terms and are believed to be of higher quality than Topic 2. Topic models are built to identify latent topics existing in a document collection. In most cases, topic models are used to gain an understanding of the collection and to find ways to categorize and characterize documents. While the method is automatic, it requires interpretable output to be useful to the modeler. In this sense, it requires a combination of art and science. To this end, human coders are oftentimes used to evaluate the topics to determine if there is a natural label that can be assigned to the topic assignments from the topic model. For this reason, Chang et al. (2009) investigate the interpretability of models compared to their quantitative performance measures. They propose the use of word intrusion and topic intrusion methods, which involve presenting

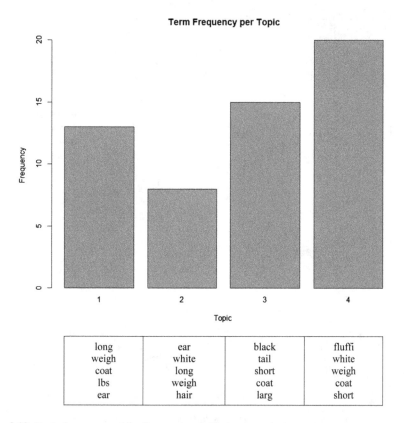

long	ear	black	fluffi
weigh	white	tail	white
coat	long	short	weigh
lbs	weigh	coat	coat
ear	hair	larg	short

Fig. 8.13 Topic frequency and the five most probable terms per topic

the most probable terms and topics and an intruder. Then, human coders are instructed to identify the intruder. Another common approach is the use of coding by field experts in the relevant domain.

Alternatively, the model can be built on the sample, with a portion removed as a holdout sample. In doing so, two measures, perplexity and held-out likelihood, can be used to assess the model. Perplexity measures how well the model predicts the held-out sample. Perplexity values that are lower are preferred and indicate that the model is a good fit. We can also compute the log-likelihood of the held-out documents. The higher the log-likelihood, the better the model fit.

8.7.3 When to Use Topic Models

When determining if topic modeling should be used in the analysis, there are several considerations to keep in mind. First, most topic model approaches are unsupervised and assume that there is uncertainty in the documents. If the true topics of the documents in the document collection are known, it may be more beneficial to apply a supervised analysis method, such as classification analysis, which is covered in Chap. 9. Topic models form soft clusters because they are typically mixed-membership models. The results of the analysis will produce the most likely topics to assign to documents and the most probable terms for each of the topics. If hard clusters are preferred and the topical content across terms and documents does not need to be considered simultaneously, cluster analysis can be used to cluster either terms or documents into hard clusters. Topic models, however, are particularly useful in making predictions. Since topic models are probabilistic models, predictions can be made about new documents based on the model.

Key Takeaways
- Topic models were created as an extension of a probabilistic version of latent semantic analysis.
- Topic models are unsupervised analysis methods that produce a topic distribution over terms and document distribution over topics.
- Popular topic modeling approaches include latent Dirichlet allocation, correlated topic models, dynamic topic models, supervised latent Dirichlet allocation, and structural topic models.

References

Arun, R., Suresh, V., Madhavan, C. V., & Murthy, M. N. (2010, June). On finding the natural number of topics with latent dirichlet allocation: Some observations. In *Pacific-Asia Conference on Knowledge Discovery and Data Mining* (pp. 391–402). Berlin/Heidelberg: Springer.

Blei, D. M. (2012). Probabilistic topic models. *Communications of the ACM, 55*(4), 77–84.

Blei, D. M., & Lafferty, J. D. (2006). Dynamic topic models. In *Proceedings of the 23rd International Conference on Machine Learning* (pp. 113–120). ACM.

Blei, D. M., & Lafferty, J. D. (2007). A correlated topic model of science. *The Annals of Applied Statistics, 1*(1), 17–35.

Blei, D. M., & Lafferty J. D. (2009). Topic models. In A. Srivastava & M. Sahami (Eds.), *Text mining: Classification, clustering, and applications.* London: Chapman & Hall/CRC Data Mining and Knowledge Discovery Series.

Blei, D. M., Ng, A. Y., & Jordan, M. I. (2002). Latent Dirichlet allocation. In *Advances in neural information processing systems* (pp. 601–608). Cambridge, MA: MIT Press.

Blei, D. M., Ng, A. Y., & Jordan, M. I. (2003). Latent Dirichlet allocation. *Journal of Machine Learning Research, 3*, 993–1022.

Blei, D., Carin, L., & Dunson, D. (2010). Probabilistic topic models. *IEEE Signal Processing Magazine, 27*(6), 55–65.

Blei, David M., & Lafferty, J.D. (2007). A Correlated Topic Model of Science. *The Annals of Applied Statistics. 1*(1): 17–35.

Cao, J., Xia, T., Li, J., & Zhang Y., & Tang, S. (2009). A density-based method for adaptive lDA model selection. *Neurocomputing — 16th European Symposium on Artificial Neural Networks 2008, 72*(7–9), 1775–1781.

Chang, J., Gerrish, S., Wang, C., Boyd-Graber, J. L., & Blei, D. M. (2009). Reading tea leaves: How humans interpret topic models. In *Advances in neural information processing systems* (pp. 288–296). Cambridge, MA: MIT Press.

Deveaud, R., SanJuan, E., & Bellot, P. (2014). Accurate and effective latent concept modeling for ad hoc information retrieval. *Document numérique, 17*(1), 61–84.

Griffiths, T. L., & Steyvers, M. (2004). Finding scientific topics. *Proceedings of the National Academy of Sciences, 101*(suppl 1), 5228–5235.

Griffiths, T. L., Steyvers, M., & Tenenbaum, J. B. (2007). Topics in semantic representation. *Psychological Review, 114*(2), 211–244.

Hofmann, T. (1999, July). Probabilistic latent semantic analysis. In *Proceedings of the Fifteenth Conference on Uncertainty in Artificial Intelligence* (pp. 289–296).

Mcauliffe, J. D., & Blei, D. M. (2008). Supervised topic models. In *Advances in neural information processing systems* (pp. 121–128). Cambridge, MA: MIT Press.

Mimno, D., Wallach, H. M., Talley, E., Leenders, M., & McCallum, A. (2011, July). Optimizing semantic coherence in topic models. In *Proceedings of the conference on empirical methods in natural language processing* (pp. 262–272). Association for Computational Linguistics.

Roberts, M. E., Stewart, B. M., Tingley, D., & Airoldi, E. M. (2013, January). The structural topic model and applied social science. In *Advances in neural information processing systems workshop on topic models: computation, application, and evaluation* (pp. 1–20).

Roberts, M., et al. (2014). Structural topic models for open-ended survey responses. *American Journal of Political Science, 58*, 1064–1082.

Yi, X., & Allan, J. (2009, April). A comparative study of utilizing topic models for information retrieval. In *European conference on information retrieval* (pp. 29–41). Berlin/Heidelberg: Springer.

Further Reading

To learn more about topic models, see Blei (2012), Blei and Lafferty (2009), and Griffiths et al. (2007).

Chapter 9
Classification Analysis: Machine Learning Applied to Text

Abstract This chapter introduces classification models. We begin with a description of the various measures for determining the model's strength. Then, we explain popular classification models including Naïve Bayes, k-nearest neighbors, support vector machines, decision trees, random forests, and neural networks. We demonstrate the use of each model with the data from the example with the four dog breeds.

Keywords Classification analysis · Categorization · Machine learning · Text classification · Text categorization · Supervised learning · Artificial neural networks · Decision trees · Random forest · Support vector machines · Naïve Bayes · k-nearest neighbors

9.1 Introduction

Think about the last time you received a spam email message in your inbox. If this experience is a distant memory, you are already familiar with classification analysis. Every time a spam email is directed to your spam or trash folder rather than your inbox, a classification model is identifying the message as spam and redirecting the message to your spam folder.

Text classification, also known as text categorization, is used extensively in text analytics. Text categorization methods have been used in information retrieval (IR) for text filtering, document organization, word-sense disambiguation, and indexing (Sebastiani 2002). More specifically, it has been used for news filtering, document organization and retrieval, opinion mining, email classification, and spam filtering (Aggarwal and Zhai 2012).

Classification analysis is a type of supervised learning model. Supervised learning models are used to automatically categorize or classify text documents when the true classification of the document is known (Kantardzic 2011; Sebastiani 2002). Classification analysis is different from cluster analysis, topic models, and latent semantic analysis because there is certainty in the natural groupings or underlying

© Springer Nature Switzerland AG 2019   131
M. Anandarajan et al., *Practical Text Analytics*, Advances in Analytics and Data Science 2, https://doi.org/10.1007/978-3-319-95663-3_9

factors of the data. In unsupervised analysis, we must infer the naturally occurring groupings in the data, without knowing the actual groupings or categories. This type of research produces inductive inference. A major advantage of classification analysis is the ability to make predictions. Classification analysis also differs from other models with respect to the analysis process and the assessment of the model fit.

9.2   The General Text Classification Process

Prior to conducting classification analysis, we split the document sample into at least two groups: the training set and the testing set. Sometimes, a third group is also included, known as the validation set. In this chapter, we perform classification analysis using a training set and a testing set. We begin with our full document set, which includes the known class labels. These labels are categorical in our example and take on discrete values. Categorical variables are qualitative labels. For instance, a document collection describing fruit may have documents with categorical class labels such as apple, banana, and orange.

We split the set into two groups based on a chosen split rule. When choosing how to split the data, there is a trade-off. We want to include enough documents in the training set to produce a good classifier and map the documents to the class label. However, we also need to have enough documents "held out" to assess the performance of our classifier. In splitting the documents, we want to include enough documents of each classification in each set, so that each class, or category, is considered when building the classifier and can be predicted. A popular split is to put 70% of the data in the training set and 30% in the testing set.

After the documents are split, the classification method is applied to the term-document matrix (TDM) or document-term matrix (DTM) of the training set. We take this step to map the relationship between the TDM or DTM and the class label. Once the classifier is created, it is applied to the TDM or DTM of the testing set, which does not include the class label. This step produces either a most likely category or a category to which the document is predicted to belong. Then, measures of external validity can assess the performance of the analysis. Figure 9.1 provides an overview of the classification analysis process (Dorre et al. 1999).

9.3   Evaluating Model Fit

9.3.1   Confusion Matrices/Contingency Tables

A contingency table, or confusion matrix, presents the actual and predicted classes based on classification analysis models. An example of a simple 2-by-2 table with two classifications, Yes and No, appears in Fig. 9.2. The cells in the table contain frequency values for each of the classifications present in the model across the

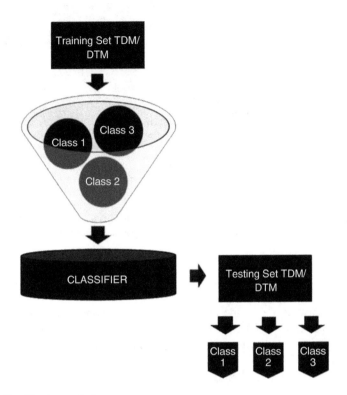

Fig. 9.1 Classification analysis process

		Actual	
		Yes	*No*
Predicted	*Yes*		
	No		

Fig. 9.2 Sample contingency table with two classifications, Yes and No

actual and predicted classifications. The cells in which the actual and predicted values match are on the diagonal, highlighted in green. These green cells represent *hits* or instances where the classification model makes a correct prediction. The cells in the contingency table where the classification model predicts the wrong classification are shown in red. These red cells represent misclassifications or *misses*.

Contingency tables allow for the efficient calculation of measures of goodness of fit. To demonstrate, we use the contingency table displayed in Fig. 9.3, which is based on Fig. 9.2. An additional row and column have been added to the table to display the total actual and predicted values, respectively. As the table demonstrates, the size of the testing data is 20, with 10 classified as Yes and 10 classified as No.

		Actual		
		Yes	*No*	**Total Predicted**
Predicted	*Yes*	3	6	9
	No	7	4	11
	Total Actual	10	10	20

Fig. 9.3 Contingency table example

9.3.2 *Overall Model Measures*

We use the contingency table to perform calculations to assess our classification analysis model fit. The first overall model measure is accuracy.

9.3.2.1 Accuracy

Accuracy measures the proportion of correct predictions. Accuracy is usually expressed as a percentage. In calculating accuracy from a contingency table, we find correctly predicted classes along the diagonal of the contingency table. We measure accuracy as

$$\text{Accuracy} = \frac{\#\,\text{of correct predictions}}{\#\,\text{of total predictions}} * 100$$

We can calculate the accuracy of the classification depicted in Fig. 9.3 as

$$\text{Accuracy} = \frac{3+4}{20} * 100 = 35\%$$

This result is very low level of accuracy. An accuracy level equal to 100% means that the model has perfect predictive power, whereas accuracy equal to 0% means that the model does not make any correct predictions.

9.3.2.2 Error Rate

The error rate considers the classification predictions that are not on the diagonal. We want a classification model with a high level of accuracy and a low error rate. Since accuracy is measured as a percentage, we calculate the error rate by subtracting the accuracy from 100. The error rate is calculated as

$$\text{Error Rate} = \frac{\#\,\text{of incorrect predictions}}{\#\,\text{of total predictions}} * 100 = \left(100 - \text{Accuracy}\right)$$

To calculate the error rate of the classification model represented in the contingency table in Fig. 9.3, we use this formula:

$$\text{Error Rate} = \frac{7+6}{20} * 100 = 65\%$$

Accuracy and error rate are two simple measures that tell us the overall fit of the model and its performance. Additional measures of the predictive power of the classification model include the Rand index, the Adjusted Rand index (ARI), and Cohen's kappa. The ARI is used to determine the agreement between the actual and predicted classifications that are not attributed to chance. Cohen's kappa considers the accuracy of the model versus its expected accuracy.

9.3.3 Class-Specific Measures

Class-specific measures reveal if our model is good (or bad) at predicting a certain class. In cases where it is more important to identify a specific class, these measures help the analyst choose the model. For instance, suppose we have a document collection of customer reviews for a subscription service. The documents are labeled "Active" if the customer is a current customer, "3–5 months" if the customer left the service after 3–5 months, and "6–12 months" if the customer left the service after 6–12 months. If we want a model that can be applied to a group of new customers for targeted marketing to those customers who are likely to leave after 3–5 months, it is imperative that the model correctly predicts this class of customers. The three class-level measures of predictive performance that we use are precision, recall, and F-measure.

9.3.3.1 Precision

Precision is the class-specific equivalent of accuracy. Precision measures how many of the predictions for a given class are accurate compared to the total number of documents that are predicted in that class. Precision does not consider the actual number of documents in the class, only the number of predictions. Precision is measured as

$$\text{Precision}_i = \text{Accuracy}_i = \frac{\# \text{ of correct predicted}_i}{\text{total} \# \text{ predicted}_i}$$

We calculate the precision of the two classes, Yes and No, of the classification model represented in the contingency table in Fig. 9.3 as

$$\text{Precision}_{\text{YES}} = \frac{3}{9} = 0.33 \text{ and Precision}_{\text{NO}} = \frac{4}{11} = 0.36.$$

Consistent with our model-level accuracy, the result is a low level of precision for the two categories.

9.3.3.2 Recall

An alternative measure to precision is recall. Recall measures how many of the predictions for a given class are correct compared to the total number of documents that truly belong to that class. Recall is measured as

$$\text{Recall}_i = \frac{\#\text{ of correct predicted}_i}{\text{total}\#\text{ of actual}_i}$$

We calculate the recall of the two classes, Yes and No, of the classification model represented in the contingency table in Fig. 9.3 as

$$\text{Recall}_{\text{YES}} = \frac{3}{10} = 0.30 \text{ and Recall}_{\text{NO}} = \frac{4}{10} = 0.40$$

9.3.3.3 F-Measure

In real-world classification, we want high levels of both precision and recall. However, this outcome is not always possible. For this reason, in assessing our model fit and choosing our classification model, we want to balance the two measures. The F-measure is a goodness of fit assessment for a classification analysis that balances precision and recall. The maximum possible value of the F-measure is 1. The F-measure is calculated as

$$F_i = 2 * \frac{\text{precision} * \text{recall}}{\text{precision} + \text{recall}}$$

We can calculate the F-measure values of the two classes, Yes and No, of the classification model represented in the contingency table in Fig. 9.3 as

$$F_{\text{YES}} = 2 * \frac{0.33 * .30}{0.33 + .30} = 0.31 \text{ and } F_{\text{NO}} = 2 * \frac{0.36 * 0.40}{0.36 + 0.40} = 0.38$$

The mean class-level F-measure values can be used as the overall F-measure. In this case, our overall F-measure value is (0.38 + 0.31)/2, 0.35. An overall F-measure value of 0.35 is very low but should be expected, given the poor performance of the classification model. Next, we introduce the classification models.

9.4 Classification Models

There are many classification models available to categorize and classify documents in text analytics applications. In this chapter, we cover six popular models that are also utilized in machine learning and data mining applications: naïve Bayes, k-nearest neighbors, support vector machines, decision trees, random forests, and neural networks.

The classification examples in this chapter use the DTM based on 56 documents in which people describe the physical appearance of their pet dogs. There are four different breeds of dogs described in the documents: Bichon Frise, Dachshund, Great Dane, and Golden Retriever. The dogs vary in terms of height, weight, size, color, and fur. Each of these breeds has distinguishing features that are characteristic of that breed. For instance, Bichon Frises are small, fluffy dogs that are predominantly white and usually have black noses. On the other hand, a Great Dane is a very large dog with a short, straight coat that can be one of several colors or a mix of colors. In the document sample, each dog type has 14 documents.

Pre-processing and parsing is applied to the 56 text documents, including the removal of stop words and stemming. Since some of the pet owners describe their dogs in terms of weight, we do not remove numbers from our corpus. The *tfidf* weighted DTM is used, and the resulting vocabulary contains 96 terms. For demonstrative purposes, we put 70% of our observations in our training set and 30% in our testing set. In this example, we keep an equal number of observations per classification in each set. There are 40 documents in the training data, 10 per dog, and 16 documents in the testing data, 4 per dog.

9.4.1 Naïve Bayes

Naïve Bayes (NB) models are often applied in the identification of spam email messages and for fraud detection. NB models are widely used because they are easy to implement and simple to use (Allahyari et al. 2017; Kim et al. 2006). The model relies on Bayes' rule, which is used to estimate conditional probability. Bayes' rule states that

$$P(X \mid Y) = \frac{P(Y \mid X) P(X)}{P(Y)}$$

NB assumes that terms are conditionally independent given the class, which simplifies the necessary calculations. The naïve Bayes classification rule says that the most likely classification $\hat{C}$, of document D, is equal to

$$\hat{C} = \arg\max P(D \mid C) * P(C),$$

Table 9.1 Naïve Bayes contingency table

		Actual			
		Bichon Frise	Dachshund	Golden Retriever	Great Dane
Predicted	Bichon Frise	4	1	0	0
	Dachshund	0	3	0	0
	Golden Retriever	0	0	4	0
	Great Dane	0	0	0	4

Table 9.2 Goodness of fit measures, naïve Bayes model

	Precision	Recall	F-measure
Bichon Frise	0.8	1.0	0.9
Dachshund	1.0	0.8	0.9
Golden Retriever	1.0	1.0	1.0
Great Dane	1.0	1.0	1.0

where $P(D|C)$ is the conditional probability of a document given its class and $P(C)$ is the probability of the classification. Since we know that documents are comprised of terms, we can determine the probability of a document given the classification by multiplying each of the probabilities of a term given that it belongs to that class. This assumption makes the model estimation efficient, but it is not necessarily accurate in applications using real-world data (Zhang 2004).

In our example, we apply the naïve Bayes model to the training data to develop the classification rule, and then the rule is used on the testing data to predict the most likely classification. Table 9.1 displays the contingency table with the most likely class labels estimated for the 16 documents in the testing set. The accurate predictions are displayed along the diagonal in green, and the misclassifications are indicated in red.

While the model is known to be simple, it is very accurate in predicting three out of the four dog breeds. The model misclassifies only one document, classifying it as Bichon Frise, when it actually describes a Dachshund. The overall accuracy of the model is 93.8%, and the overall F-measure is 0.95. The class-level goodness of fit measures, precision, recall, and F-measure are displayed in Table 9.2.

9.4.2   k-Nearest Neighbors (kNN)

Similar to the distance-based groupings formed in clustering, in k-nearest neighbor (kNN) classifications, the similarity between documents is used to classify them (Aggarwal and Zhai 2012). The most common classification in a document's k-nearest neighbor group is set as that group's class label. The training groups are then compared to the testing points, and the group nearest to each testing point becomes that point's predicted classification. kNN is considered a "lazy learning model," meaning that no actual model is created. Instead, all calculations are performed during the classification of the testing data (Bell 2006).

Table 9.3 1NN testing document actual classifications, 1NN documents and 1NN predicted classifications

Testing Document	Testing Classification	1NN Document	1NN Predicted Classification
2	BF	15	GR
6	BF	14	BF
8	BF	4	BF
9	BF	4	BF
15	GR	21	GR
17	GR	22	GR
20	GR	21	GR
27	GR	22	GR
31	GD	21	GR
32	GD	34	GD
36	GD	16	GR
38	GD	40	GD
43	D	3	BF
48	D	35	GD
52	D	46	D
53	D	46	D

Table 9.4 Contingency table, kNN classification, $k = 1$ (1NN)

		Actual			
		Bichon Frise	Dachshund	Golden Retriever	Great Dane
Predicted	Bichon Frise	3	1	0	0
	Dachshund	0	2	0	0
	Golden Retriever	1	0	4	2
	Great Dane	0	1	0	2

To illustrate the use of the kNN classification model, let's consider an example using $k = 1$. In this case, a distance matrix of the DTM is used to find the closest data point in the training data for each of the testing data points. Based on the distance values in the distance matrix, the nearest neighbors to the 16 testing documents and their resulting most likely classifications are shown in Table 9.3. Misclassified testing documents are shown in red, and correctly classified documents are shown in green.

The resulting contingency table appears in Table 9.4 and displays the correct classifications in green and the misclassifications in red. The kNN model with $k = 1$, or 1NN model, correctly predicts the documents about Golden Retrievers but has trouble predicting the other classes. The overall accuracy of the model is 68.8%.

The class-level goodness of fit measures, precision, recall, and F-measure are displayed in Table 9.5. While the model has perfect precision for the Dachshund class, it has very poor recall (0.5). The model has perfect recall for the Golden Retriever class, but only a 0.6 precision level. The class with the highest F-measure value is Bichon Frise, with precision and recall values of 0.8.

Table 9.5 Goodness of fit measures, k-nearest neighbors, $k = 1$

	Precision	Recall	F-Measure
Bichon Frise	0.8	0.8	0.8
Dachshund	1.0	0.5	0.7
Golden Retriever	0.6	1.0	0.7
Great Dane	0.7	0.5	0.6

An obvious caveat of kNN classification is that it requires a k value prior to performing the classification. Additionally, the solution is sensitive to the chosen value of k. However, since the number of classifications is known in supervised learning, the best k to start with is usually the number of classes present in the dataset. For large datasets, kNN can be very inefficient.

9.4.3 Support Vector Machines (SVM)

Support vector machine (SVM) classification involves the search for the optimal hyperplane, among all possible separating hyperplanes, which has the largest margin of separation between classes (Hearst et al. 1998; Vapnik and Kotz 1982). The support vectors are the points located at the boundary of the margin. According to Joachims (1998), SVM performs very well when applied to text classification because SVM can handle the high dimensionality and sparsity of the DTM used as the input quite efficiently (Joachims 1998).

An illustration of a two-dimensional SVM classification model is displayed in Fig. 9.4. The blue circles representing Documents 1–5 belong to class "blue" and the green circles representing Documents 6–10 belong to class "green." The optimal hyperplane is the red line depicted in the figure. The red line is the optimal hyperplane because it creates the largest margin, shown in purple, between the two support vectors, which are represented as orange dashed lines drawn through Documents 4 and 10.

We apply SVM to our example DTM to predict the four dog breeds. The resulting contingency table is shown in Table 9.6, which displays the correct classifications in green and the misclassifications in red. The overall accuracy of the SVM model is 81.3%.

The goodness of fit measures, precision, recall, and F-measure are displayed in Table 9.7. Overall, the model performs very well in predicting the class labels. The SVM model has perfect precision for the Bichon Frise and Dachshund breed classes, and perfect recall for the Golden Retriever class. All of the classes have high F-measure values over 0.8, and the overall F-measure is 0.85.

Overall, SVM is accurate, efficient, and quick to train and has been shown to outperform other methods, including naïve Bayes and the next method that we will cover: decision trees (Dumais et al. 1998). On the other hand, Ikonomakis et al. (2005) assert that despite the high level of precision of the SVM model, it can suffer from poor recall.

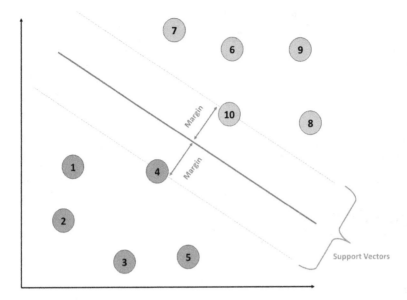

Fig. 9.4 Two-dimensional representation of support vector machine classification of ten documents (Sebastiani 2002)

Table 9.6 Support vector machines contingency table

		Actual			
		Bichon Frise	Dachshund	Golden Retriever	Great Dane
Predicted	Bichon Frise	3	0	0	0
	Dachshund	0	3	0	0
	Golden Retriever	0	1	4	1
	Great Dane	1	0	0	3

9.4.4   Decision Trees

Decision tree classification is a nonparametric approach that uses recursive partitioning to separate classes within a dataset (Quinlan 1986; Rokach and Maimon 2005; Sebastiani 2002). Decision trees are comprised of a collection of rules dividing a dataset into successively smaller groups for classification or prediction. Decision trees help the user understand and explain groups and subgroups occurring in the data, because they are based on a series of rules. The output of using DT is a visualization depicting the partitioning process.

The top node leading to all subsequent partitions is known as the root node. The goal of each split in a decision tree is to increase purity, which refers to the homogeneity of the partitioned data. With each successive split, we want to improve the groups so that each group includes a single class. A pure node would contain all points of a single class and would have a purity value of 1. There are

Table 9.7 Goodness of fit measures, SVM

	Precision	Recall	F-Measure
Bichon Frise	1.0	0.8	0.9
Dachshund	1.0	0.8	0.9
Golden Retriever	0.7	1.0	0.8
Great Dane	0.8	0.8	0.8

Gini Index	*measures divergence between probability distributions (Rokach and Maimon 2005)*
• Purer nodes have a value of 1; therefore, higher Gini Index values are preferred when making splitting decisions (Berry and Linoff 2011). • The Classification and Regression Tree (CART) method uses Gini (Breiman et al. 1984).	

Entropy/Deviance	*measures homogeneity*
• Purer nodes have a value of 0; therefore, lower Entropy values are preferred when making splitting decisions (Sutton 2005).	

Information Gain	*measures the reduction in Entropy*
• Higher values are preferred when making splitting decisions.	

Chi-Square Test	*measures the likelihood of a split*
• A higher value indicates that the split is less likely to be due to chance and is a better split than a split with a lower value (Loh 2008). • Increasing in the dataset size.	

Fig. 9.5 Splitting criteria for decision trees

several measures of purity to split the nodes of a decision tree, including Gini, entropy, chi-square, and deviance. Figure 9.5 describes the four purity measures.

Using the training set, we utilize entropy as the splitting criteria to create a classification tree. The resulting tree, shown in Fig. 9.6, is formed using four terms: *white*, *short*, *black*, and *ear*, with *white* as the root node.

The resulting contingency table is shown in Table 9.8, which displays the correct classifications in green and the misclassifications in red. The overall prediction accuracy of the DT classification model is 62.5%. As the table indicates, the model is particularly poor at predicting the Great Dane breed category.

The class-level goodness of fit measures, precision, recall, and F-measure are displayed in Table 9.9. While the model has perfect recall for the Bichon Frise class and perfect precision for the Golden Retriever class, the model is very poor at predicting the documents about Great Danes, with both precision and recall values of 0.3. The overall F-measure value is 0.63.

Decision trees used for classification have many beneficial properties. They are not sensitive to outliers or distributional properties. They are also simple and easy to understand. The visual aid adds concreteness to the results of the analysis.

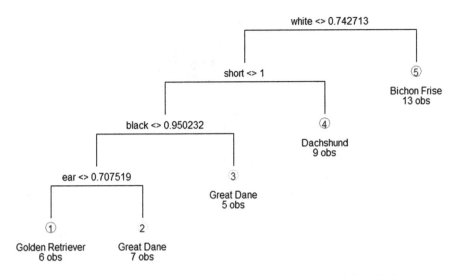

Fig. 9.6 Decision tree created from training data using deviance as the splitting criteria

Table 9.8 Decision tree confusion matrix

		Actual			
		Bichon Frise	Dachshund	Golden Retriever	Great Dane
	Bichon Frise	4	1	0	2
Predicted	Dachshund	0	2	0	1
	Golden Retriever	0	0	3	0
	Great Dane	0	1	1	1

Table 9.9 Goodness of fit measures, decision tree

	Precision	Recall	F-Measure
Bichon Frise	0.6	1.0	0.7
Dachshund	0.7	0.5	0.6
Golden Retriever	1.0	0.8	0.9
Great Dane	0.3	0.3	0.3

However, decision trees can be prone to overfitting, where the model demonstrates strong predictive performance on the training set but may fail to be generalizable. For this reason, decision trees can be sensitive to minor changes in the training set. In addition, they are computationally expensive (Sutton 2005).

9.4.5 Random Forests

Random forests (RF) are a "combination of tree predictors such that each tree depends on the values of a random vector sampled independently and with the same distribution for all trees in the forest" (Breiman 2001, 1). RF can overcome the

Table 9.10 Random forest contingency table

		Actual			
		Bichon Frise	Dachshund	Golden Retriever	Great Dane
Predicted	Bichon Frise	4	0	0	1
	Dachshund	0	3	0	0
	Golden Retriever	0	0	4	0
	Great Dane	0	1	0	3

Table 9.11 Goodness of fit measures, random forest

	Precision	Recall	F-Measure
Bichon Frise	0.8	1.0	0.9
Dachshund	1.0	0.8	0.9
Golden Retriever	1.0	1.0	1.0
Great Dane	0.8	0.8	0.8

instability of a single decision tree by using many different decision trees, creating a forest of decision trees that on average are more accurate (Strobl et al. 2009). The procedure involves building many trees that form a collection of trees, or a forest. From that forest, voting or averaging can be used to make classification decisions.

We apply RF to the DTM example to predict the dog breed classification. The resulting contingency table is shown in Table 9.10, which displays the correct classifications in green and the misclassifications in red. The overall accuracy is 87.5%.

The class-level goodness of fit measures, precision, recall, and F-measure are displayed in Table 9.11. The RF model clearly outperforms the DT model. In fact, the model is able to predict the Golden Retriever class with perfect precision and recall. The model also has perfect precision for the Dachshund class and perfect recall for the Bichon Frise class. The F-measure values are all at or above 0.8, and the overall F-measure value is 0.9.

Based on the contingency table in Table 9.10 and the fit measures in Table 9.11 for the RF predictions, it is clear that the use of many decision trees in RF outperforms the use of a single DT across all fit measures. One feature of RF analysis is the ability to measure and visualize the importance of variables based on the analysis. Fig. 9.7 displays the importance of variables based on the mean decrease in accuracy. The importance of the terms *short* and *white* is evident in this figure. RF has the advantage of accuracy and efficiency and overcomes the overfitting issue of a classification model with a single DT (Breiman 2001).

9.4.6   Neural Networks

Neural networks (NN) or artificial neural networks (ANN) are made up of interconnected groups of nodes with one or more hidden layers linking inputs to outputs and are based on the neural networks that exist in the human brain (Kantardzic 2011).

Fig. 9.7 Random forest plot of the importance of variables

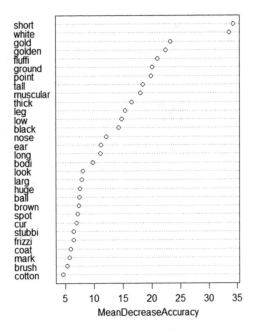

NN is a machine learning technique that learns and adapts based on the data in the model. Neural networks are referred to as a black box model because they have a high level of accuracy but are difficult to understand or interpret. For this reason, this network-based approach is best when we want to create a high-performing classification model, but an explanation of the model is not crucial to the application of the model. The model requires that the number of hidden layers and hidden layer nodes be specified in advance.

Figure 9.8 depicts an example of NN. A NN classification model takes the model input and maps the input to the output through a series of nodes and one or more hidden layers. In the figure, nodes are depicted as blue circles. There are five input nodes in the input layer, one for each input. There is one hidden layer with four nodes, which connect to the three output nodes in the output layer.

In the NN classification model predicting dog breed, we use one hidden layer with five hidden nodes. The resulting contingency table is shown in Table 9.12, which displays the correct classifications in green and the misclassifications in red. The overall accuracy of the model is 75%.

The goodness of fit measures, precision, recall, and F-measure are displayed in Table 9.13. The NN model has perfect precision for the Bichon Frise class and perfect recall for the Golden Retriever and Great Dane classes. However, based on the F-measure, which balances the two performance measures, the model does not perform well for the Bichon Frise and Dachshund classes. On the other hand, the NN classification is quite successful in predicting the Golden Retriever and Great Dane classes. The overall F-measure value of the NN classification model is 0.75.

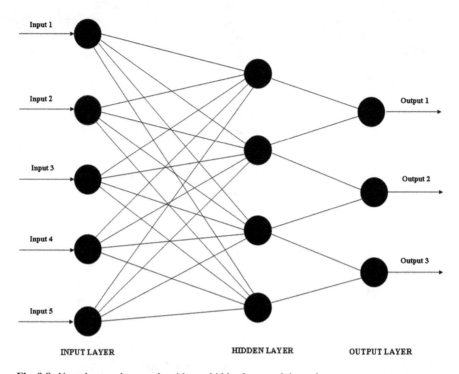

Fig. 9.8   Neural network example with one hidden layer and three classes

Table 9.12   Neural network contingency matrix with five hidden nodes in one hidden layer

		Actual			
		Bichon Frise	**Dachshund**	**Golden Retriever**	**Great Dane**
Predicted	Bichon Frise	2	0	0	0
	Dachshund	1	2	0	0
	Golden Retriever	0	2	4	0
	Great Dane	1	0	0	4

9.5   Choosing a Classification

9.5.1   Model Fit

There are many ways to evaluate which model to choose for text classification. One way is to assess the accuracy of the model. In the models we created here, the most accurate model is the simplest: naïve Bayes. The overall accuracy measures for each model, in descending order of accuracy, are presented in Table 9.14. As the table illustrates, the most accurate models are Naïve Bayes and random forest, while the least accurate are *k*-nearest neighbors and decision trees. Other measures of model

Table 9.13 Goodness of fit measures, neural network with five hidden nodes in one hidden layer

	Precision	Recall	F-measure
Bichon Frise	1.0	0.5	0.7
Dachshund	0.7	0.5	0.6
Golden Retriever	0.7	1.0	0.8
Great Dane	0.8	1.0	0.9

Table 9.14 Classification model accuracy

Model	Accuracy (%)
Naïve Bayes	94
Random forest	89
Support vector machines	81
Neural networks	75
k-nearest neighbors	69
Decision tree	63

fit, as described in the beginning of this chapter, can also be used to consider the performance of the classification model in predicting the document class.

In choosing the classification analysis method, it is important to consider the strengths and weaknesses of the model and the specific application. If it is important not only to classify the documents but also to know the importance of specific terms in the model's creation, a decision tree or random forest classification analysis can be used. If the application requires a visual aid demonstrating how the classification is done, decision trees may be the most appropriate. On the other hand, if an understanding of the specifics of the classifier is not important to the classification application, neural network analysis can be used.

If the DTM or TDM being used in the classification analysis is very large and very sparse, one should consider the advantage of the support vector machine model's strength in classifying large, sparse data. If there is reason to believe that there are very strong naturally occurring groups of documents in the data, the k-nearest neighbors model should be a strong contender. Finally, if the application calls for simplicity and efficiency, naïve Bayes might be the best choice. Ultimately, the choice of the appropriate model depends on the priorities of the analyst: overall accuracy, precision, recall, the F-measure, or an alternate goodness-of-fit measure.

Key Takeaways
- Classification analysis applied to text uses category labeled documents as input to build a predictive model.
- Some of the machine learning techniques used for classification analysis presented in this chapter include naïve Bayes, k-nearest neighbors, decision trees, random forest, support vector machines, and neural networks.

(continued)

(continued)
- In classification analysis, a training and testing set are used. The model is built using the training sample, and the predictive performance is assessed using the testing sample.
- Measures of overall predictive performance include accuracy and error rate.
- Measure of class-specific predictive performance includes precision, recall, and F-measure.

References

Aggarwal, C. C., & Zhai, C. X. (2012). *Mining text data*. New York: Springer-Verlag.

Allahyari, M., Pouriyeh, S., Assefi, M., Safaei, S., Trippe, E. D., Gutierrez, J. B., & Kochut, K. (2017). A brief survey of text mining: Classification, clustering and extraction techniques. arXiv preprint arXiv:1707.02919.

Bell, D. A. (2006). Using kNN model-based approach for automatic text categorization. *Soft Computing, 10*(5), 423–430.

Berry, M. J., & Linoff, G. S. (2011). *Data mining techniques. For marketing, sales, and customer relationship management*. Chichester: Wiley-Interscience.

Breiman, L. (2001). Random forests. *Machine learning, 45*(1), 5–32.

Dörre, J., Gerstl, P., & Seiffert, R. (1999). Text mining: Finding nuggets in mountains of textual data. In *Proceedings of the fifth ACM SIGKDD International Conference on Knowledge Discovery and Data Mining* (pp. 398–401). ACM.

Dumais, S., Platt, J., Heckerman, D., & Sahami, M. (1998, November). Inductive learning algorithms and representations for text categorization. In *Proceedings of the Seventh International Conference on Information and Knowledge Management* (pp. 148–155). ACM.

Hearst, M. A., Dumais, S. T., Osuna, E., Platt, J., & Scholkopf, B. (1998). Support vector machines. *IEEE Intelligent Systems and Their Applications, 13*(4), 18–28.

Ikonomakis, M., Kotsiantis, S., & Tampakas, V. (2005). Text classification using machine learning techniques. *WSEAS Transactions on Computers, 4*(8), 966–974.

Joachims, T. (1998). Text categorization with support vector machines: Learning with many relevant features. In *Machine learning: ECML-98* (pp. 137–142). Berlin/Heidelberg: Springer.

Kantardzic, M. (2011). *Data mining: Concepts, models, methods, and algorithms*. New York: Wiley.

Kim, S. B., Han, K. S., Rim, H. C., & Myaeng, S. H. (2006). Some effective techniques for naive bayes text classification. *IEEE Transactions on Knowledge and Data Engineering, 18*(11), 1457–1466.

Quinlan, J. R. (1986). Induction of decision trees. *Machine Learning, 1*(1), 81–106.

Rokach, L., & Maimon, O. (2005). Top-down induction of decision trees classifiers-a survey. *IEEE Transactions on Systems, Man, and Cybernetics, Part C (Applications and Reviews), 35*(4), 476–487.

Sebastiani, F. (2002). Machine learning in automated text categorization. *ACM Computing Surveys (CSUR), 34*(1), 1–47.

Strobl, C., Malley, J., & Tutz, G. (2009). An introduction to recursive partitioning: Rationale, application, and characteristics of classification and regression trees, bagging, and random forests. *Psychological Methods, 14*(4), 323.

Sutton, C. D. (2005). Classification and regression trees, bagging, and boosting. In *Handbook of Statistics* (Vol. 24, pp. 303–329). Amsterdam: Elsevier.

Vapnik, V. N., & Kotz, S. (1982). *Estimation of dependences based on empirical data* (Vol. 40). New York: Springer-Verlag.

Zhang, T. (2004). Statistical behavior and consistency of classification methods based on convex risk minimization. *Annals of Statistics*, 56-85.

Further Reading

For more about machine learning techniques, see Berry and Linoff (2011) and Kantardzic (2011). For machine learning techniques for text classification, see Sebastiani (2002), Ikonomakis et al. (2005), and Aggarwal and Zhai (2012).

Chapter 10
Modeling Text Sentiment: Learning and Lexicon Models

Abstract This chapter presents two types of sentiment analysis: lexicon-based and learning-based. Both methods aim to extract the overall feeling or opinion from text. Each approach is described with an example, and then the difficulties of sentiment analysis are discussed.

Keywords Sentiment analysis · Opinion mining · Learning · Lexicon

Sentiment analysis and opinion mining were introduced in the early 2000s as methods to understand and analyze opinions and feelings (Dave et al. 2003; Liu 2012; Nasukawa and Yi 2003). Sentiment analysis and opinion mining can be used interchangeably, as both analyze text to understand feeling. The goal of sentiment analysis is to explain whether the document contains positive or negative emotions. Sentiment analysis has many practical applications, especially for businesses. It helps businesses determine public opinion about the brand, the effect of publicity, or the reactions to product releases (Mullich 2013).

Human readers use their knowledge of language to determine the feelings behind a text. This applies to individual words as well as their meaning within the document's context. A computer, unfortunately, needs to be told which words or phrases change the polarity of a text (Silge and Robinson 2016, 2017). In sentiment analysis, the unit of analysis can range from the document level to the word level, as depicted in Fig. 10.1 (Appel et al. 2015; Kumar and Sebastian 2012).

At the document level, the sentiment of the full document is captured to determine the overall sentiment of each individual document. At the sentence level, the same determination occurs, but each sentence is considered individually when calculating sentiment. Feature-level sentiment is measured on the attribute level, particularly when applying sentiment analysis to customer or product feedback and reviews (Appel et al. 2015).

For instance, in a customer review about a car, a feature-level analysis will consider the sentiment of the customer with respect to the look, fuel efficiency, comfort, durability, and price of the vehicle separately. While people may have negative feelings about the price and fuel efficiency of their luxury automobile, they may be more positive with respect to the look, comfort, and durability of their high-priced

© Springer Nature Switzerland AG 2019
M. Anandarajan et al., *Practical Text Analytics*, Advances in Analytics and Data
Science 2, https://doi.org/10.1007/978-3-319-95663-3_10

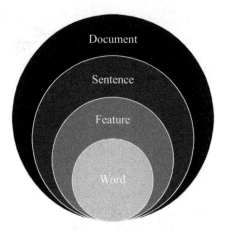

Fig. 10.1 Levels of sentiment Analysis

vehicle. Analysis at this level is more granular than at the sentence level, but less granular than analysis conducted at the term level.

At the word level, sentiment analysis typically centers on adjectives as the descriptive words with sentiment attached (Appel et al. 2015). According to Kumar and Sebastian (2012), analysis conducted at the word level can follow one of two methods: dictionary or corpus-based. Dictionary-based methods use pre-made dictionaries containing terms and sentiment scores to assign sentiment scores to text data by comparing the terms in the sample to the particular dictionary. Corpus-based, word-level sentiment analysis builds on terms in the corpus with identifiable sentiments to create an association network for synonymous terms, which can be used to classify terms with unknown sentiment (Kumar and Sebastian 2012). Next, we will explore these two methods.

There are two common approaches to sentiment analysis: dictionary, or lexicon, and corpus, or learning. The lexicon approach assigns a polarity to words from a previously created dictionary. This dictionary defines a word and its polarity. If the lexicon contains the same word or phrase that appears in the text, its polarity value is returned. In contrast, the learning-based method builds an automatic sentiment classifier for a document set previously annotated with sentiments. From there, a classifier is trained that can be applied to new, unseen data (Ignatow and Mihalcea 2016).

10.1   Lexicon Approach

The lexicon approach uses previously scored words and word phrases to assign a sentiment value to a new text. Each word or phrase that matches the corresponding word or phrase in the lexicon is given that value. For the full text, the values are then summed (Silge and Robinson 2016, 2017).

Numerous scored lexicons exist for use in sentiment analysis. Labels typically include an indicator for positive and negative or a score that indicates the strength of the polarity. Popular lexicons include OpinionFinder (Wilson et al. 2005), General Inquirer (Stone et al. 1966), SentiWordNet (Baccianella et al. 2010), and AFINN (Nielsen 2011). Of these four popular lexicons, OpinionFinder and General Inquirer provide categorical sentiment values, such as positive and negative, and SentiWordNet and AFINN provide numerical sentiment values.

OpinionFinder contains 8,223 word entries, including POS tags, stemming indicators, and sentiment classifications.[1] Sentiment classifications in OpinionFinder can be negative, neutral, positive, or both (Wilson et al. 2005). General Inquirer contains many classification category labels for terms originating from four different sources, which extend far beyond sentiment.[2]

SentiWordNet is a lexicon that uses WordNet, a lexical database, to assign numeric sentiment score scales ranging from -1 for negative sentiment tokens to $+1$ for positive sentiment tokens (Esuli and Sebastiani 2007). SentiWordNet includes words and phrases, or n-grams, with the largest phrases containing six terms.[3] The AFINN lexicon scores words in a range from -5 to 5, with negative numbers indicating negative polarity and positive numbers representing positive polarity.[4] For both of the lexicons providing numerical sentiment values, the absolute value of the sentiment score indicates the strength of its polarity. Figure 10.2 displays several positive and negative terms that appear in and have sentiment agreement across the four lexicons.

To demonstrate, we use the four lexicons to score the review presented in Fig. 10.3 at the document, sentence, and word level. As shown in the figure, the review is unitized, tokenized, standardized, and cleaned. We remove stop words after comparing stop word lists with the lexicons to avoid losing any emotive words and lemmatize the tokens.

The sentiment scores by word are displayed in Table 10.1. By calculating the sentiment at the word level, we can assess the word-, sentence-, and document-level sentiment for the review for each of the four sentiment lexicons. The terms in the first sentence are highlighted in blue, and the terms in the second sentence are highlighted in pink. Although there is disagreement across the lexicons, at the document level, there is agreement that this document is positive. At the document level, the SentiWordNet sentiment score is 0.375. The sentiment score based on AFINN is 6, which is found by summing the word-level scores. At the word level,

---

[1] The OpinionFinder subjectivity lexicon can be downloaded at http://mpqa.cs.pitt.edu/lexicons/subj_lexicon/

[2] Information about the four category sources and the General Inquirer lexicon can be downloaded at http://www.wjh.harvard.edu/~inquirer/homecat.htm

[3] The SentiWordNet lexicon can be downloaded at http://www.wjh.harvard.edu/~inquirer/homecat.htm. To learn more about WordNet, visit: https://wordnet.princeton.edu/

[4] The AFINN lexicon can be downloaded at http://www2.imm.dtu.dk/pubdb/views/publication_details.php?id=6010

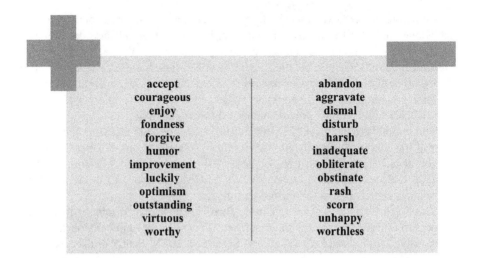

accept	abandon
courageous	aggravate
enjoy	dismal
fondness	disturb
forgive	harsh
humor	inadequate
improvement	obliterate
luckily	obstinate
optimism	rash
outstanding	scorn
virtuous	unhappy
worthy	worthless

Fig. 10.2 Sample of positive and negative words that coincide and are consistent across the four lexicons

the words *great*, *perfectly*, and *expert* are positive words, although inconsistently across the lexicons. At the sentence level, there is agreement across the four lexicons that the second sentence is positive. However, only three of the four lexicons identify the first sentence as positive, with SentiWordNet indicating that this sentence is neutral.

Next, let's try the same exercise with a negative sentence. Again, we use the four methods to score the negative review presented in Fig. 10.4. Since this review is only one sentence, the document-level and sentence-level results are identical, and we analyze at the document and word level. We follow the same preparation and preprocessing steps as in the positive review.

The sentiment scores by word for the negative review are displayed in Table 10.2. Although there is disagreement across the lexicons with respect to specific words, the lexicons agree that the document-level sentiment of the review is negative. SentiWordNet assigns an overall sentiment score of −2.13 to the review, and the score according to AFINN is −6. OpinionFinder identifies three negative terms, while General Inquirer identifies two. At the word level, the four lexicons agree that the word *ugly* is negative. Three lexicons identify *terribly* as negative, two identify *rude*, and *smell* and *loud* are listed as negative by one lexicon each.

Let's take a closer look at the AFINN results. The negative review analyzed using this lexicon returned a −6 indicating a negative polarity. Note that *rude* and *loud* did not return scores. As human beings, we would describe these words in the context of a restaurant as negative opinions. However, these words do not exist in the lexicon, so they do not get scored. Such situations demonstrate the imperfections of general-purpose sentiment analysis lexicons. In some contexts, they have a negative polarity, but that is not always true.

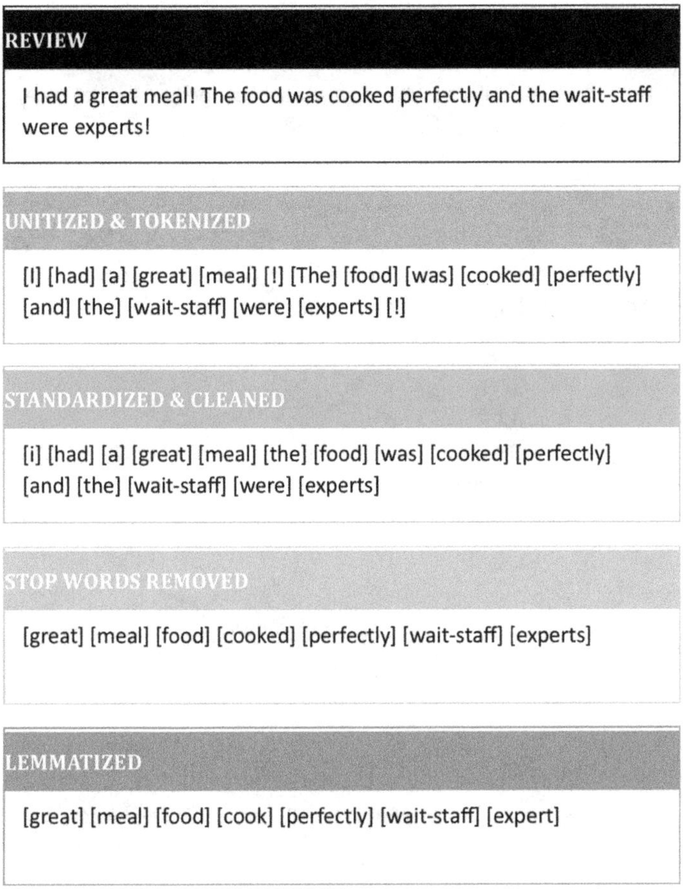

REVIEW

I had a great meal! The food was cooked perfectly and the wait-staff were experts!

UNITIZED & TOKENIZED

[I] [had] [a] [great] [meal] [!] [The] [food] [was] [cooked] [perfectly] [and] [the] [wait-staff] [were] [experts] [!]

STANDARDIZED & CLEANED

[i] [had] [a] [great] [meal] [the] [food] [was] [cooked] [perfectly] [and] [the] [wait-staff] [were] [experts]

STOP WORDS REMOVED

[great] [meal] [food] [cooked] [perfectly] [wait-staff] [experts]

LEMMATIZED

[great] [meal] [food] [cook] [perfectly] [wait-staff] [expert]

Fig. 10.3 Positive review example: text preparation and preprocessing

Table 10.1 Positive review word-, sentence-, and document-level sentiment

Word	OpinionFinder	General Inquirer	SentiWordNet	AFINN
great	+	+		3
meal				
food				
cook				
perfectly	+		0.38	3
waitstaff				
expert	+	+		
	+	+	0.38	6

REVIEW

The entrance smelled terribly, it was very loud, the hostess was rude, and the decor was ugly!

UNITIZED & TOKENIZED

[The] [entrance] [smelled] [terribly] [,] [it] [was] [very] [loud] [,] [the] [hostess] [was] [rude] [,] [and] [the] [decor] [was] [ugly] [!]

STANDARDIZED & CLEANED

[the] [entrance] [smelled] [terribly] [it] [was] [very] [loud] [the] [hostess] [was] [rude] [and] [the] [decor] [was] [ugly]

STOP WORDS REMOVED

[entrance] [smelled] [terribly] [very] [loud] [hostess] [rude] [decor] [ugly]

LEMMATIZED

[entrance] [smell] [terribly] [very] [loud] [hostess] [rude] [decor] [ugly]

Fig. 10.4 Negative review example: text preparation and preprocessing

Table 10.2 Negative review word- and document-level sentiment

	OpinionFinder	General Inquirer	SentiWordNet	AFINN
entrance				
smell			**−0.88**	
terribly	−		**−0.25**	**−3**
very				
loud			**−0.50**	
hostess				
rude	−	−		
decor				
ugly	−	−	**−0.50**	**−3**
	−	−	**−2.13**	**−6**

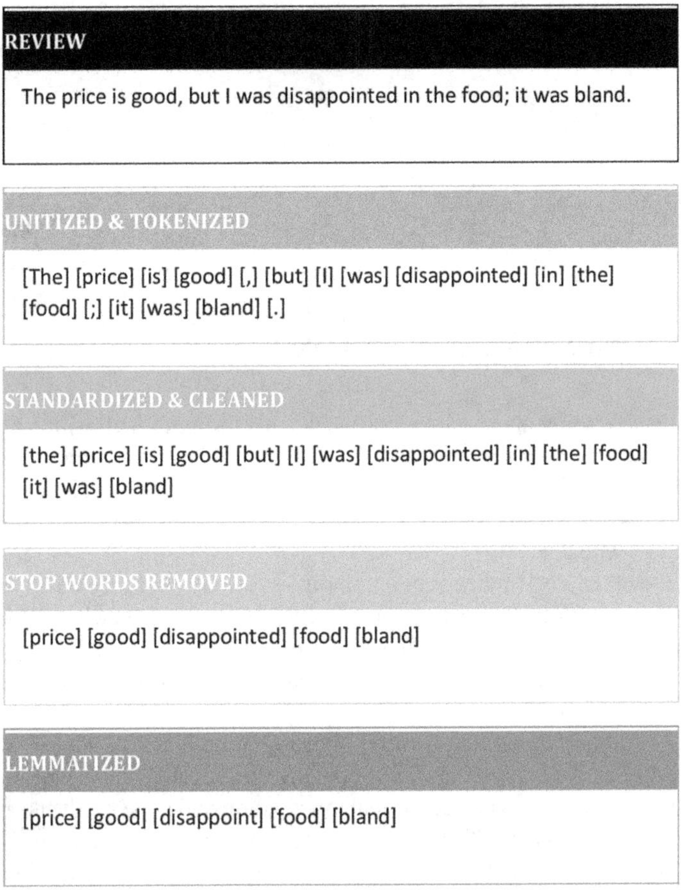

Fig. 10.5 Ambiguous review example: text preparation and preprocessing

Both of those reviews were straightforward—obviously positive or negative. Next, let's consider what happens with an ambiguous review. The prepared and preprocessed ambiguous review is presented in Fig. 10.5.

The results of sentiment analysis applied to this ambiguous review, as displayed in Table 10.3, are less consistent across lexicons than the first two examples. OpinionFinder, General Inquirer, and SentiWordNet label this review as negative, while AFINN classifies the review as positive. Unlike the other three lexicons, AFINN does not identify *bland* as a sentiment word, leading to the inconsistent scoring.

The ambiguous review demonstrates some of the difficulty associated with lexicon-based sentiment analysis. Next, we will explore an alternative method, the machine learning-based approach.

Table 10.3 Ambiguous review word and document-level sentiment

	OpinionFinder	General Inquirer	SentiWordNet	AFINN
price				
good	+	+	0.47	3
disappoint	−	−	−0.50	−2
food				
bland	−	−	−0.38	
	−	−	−0.41	1

10.2 Machine Learning Approach

The machine learning approach to sentiment analysis builds a classifier, as in Chap.
9, on a dataset with labeled sentiments. Datasets from news articles, headlines,
movie reviews, and product reviews have been annotated for this purpose. Domain-
specific datasets perform best when applied to unseen data from that domain
(Ignatow and Mihalcea 2016).

First, the classifier is trained on a training set. Then, the model is applied to the
testing data to classify the text's sentiment. Some common machine learning and
classification analysis methods applied to sentiment analysis are support vector
machines (SVM), naïve Bayes (NB), *k*-nearest neighbors (kNN), and logistic
regression (Feldman 2013).

To demonstrate the machine learning approach to sentiment analysis, we will use
a sample of 100 Amazon reviews with positive or negative class labels (Dua and
Taniskidou 2017). There is no additional information about the products, aside from
the content of the reviews. The review data are prepared and preprocessed using a
small custom stop list, including the terms *the, you, is, was, this, that, for, are,* and

Positive Review Words   **Negative Review Words**

Fig. 10.6 Word clouds of positive and negative words in review sample

am. In Chap. 13, in the example of sentiment analysis using the R Software, we explore the use of custom stop word lists further. Once the DTM is created, the sentiment label indicating if the review is positive or negative is added back to the document representation.

In defining the training and testing data, we use an 80/20 split, with 80 reviews in the training set and 20 reviews in the testing set. In both the training and testing sets, half of the reviews are positive and half are negative. Figure 10.6 displays two word clouds made up of the terms for the positive and negative reviews. The size of the terms in each of the word clouds indicates the frequency of the term in the sample.

10.2.1 Naïve Bayes (NB)

We perform a NB analysis on the unweighted DTM, following the procedures outlined in Chap. 9. The resulting contingency table of the NB sentiment analysis appears in Table 10.4. As shown, the sentiment analysis using NB has 65% predictive accuracy in classifying the sentiment of the reviews in the testing sample.

To understand the results, we take a closer look at some of the accurate and inaccurate predictions based on the testing data. As shown in Fig. 10.7, the NB senti-

Table 10.4 Naïve Bayes contingency matrix classification analysis

		Actual		
		−	+	**Total**
Predicted	−	**6**	3	9
	+	4	**7**	11
	Total	10	10	20

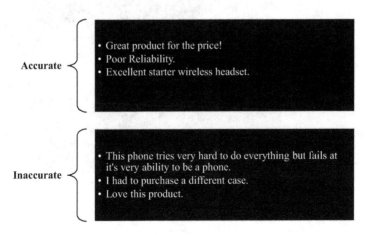

Fig. 10.7 Examples of Accurate and Inaccurate Predictions using NB

ment analysis model has difficulty classifying ambiguous reviews but performs as expected with more obvious sentiments.

10.2.2 Support Vector Machines (SVM)

We can utilize support vector machines (SVM) in the manner described in Chap. 9, using the *tfidf*-weighted DTM. The resulting contingency table of the SVM sentiment analysis appears in Table 10.5. As shown, the sentiment analysis using SVM has the same predictive power as the NB sentiment analysis model, namely, 65% predictive accuracy in classifying the sentiment of the reviews in the testing sample.

To understand the results, we take a closer look at some of the accurate and inaccurate predictions based on the testing data. As shown in Fig. 10.8, the SVM sentiment analysis model also has difficulty classifying ambiguous reviews but performs as expected with clearer sentiments.

Table 10.5 SVM contingency matrix classification analysis

		Actual		
		−	+	Total
Predicted	−	4	1	5
	+	6	9	15
	Total	10	10	20

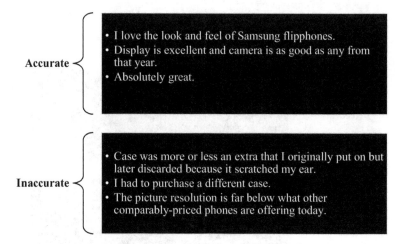

Fig. 10.8 Examples of accurate and inaccurate predictions using SVM

10.2.3 Logistic Regression

Due to the binary nature of the sentiments in our sentiment analysis, we can use a model known as logistic regression to predict review sentiment. Logistic regression is a generalized regression model for binary dependent variables, such as the positive and negative variables that we aim to predict in sentiment analysis (Kantardzic 2011). Rather than predict the category, logistic regression estimates the probability that the dependent variable will be either positive or negative. A general form of a logistic regression model is

$$\text{logit}\left(p_k\right) = \ln\left(\frac{p_k}{1-p_k}\right) = \beta_0 + \beta_1 x_{1k} + \cdots + \beta_n x_{nk},$$

where $logit(p_k)$ represents the natural log odds ratio of the outcome of interest in document k, p_k represents the probability of the classification, the β parameters are the regression coefficients for the n term variables, and the x variables are the n term weights for the k documents. In sentiment analysis using logistic regression, the independent variables are the column values in the DTM representing either the raw frequency or weighted frequency of each of the terms in the document.

For classification, logistic regression models require a chosen probability cutoff point. Anything below the cutoff is designated as one class, and the rest are categorized as the other class. Varying the cutoff point will change the classification solution. For this reason, it is beneficial to consider many potential solutions by varying the cutoff point.

We apply a logistic regression to the *tfidf*-weighted DTM and choose the cutoff to be 0.5. With this cutoff, anything at or below 0.5 is labeled as negative, and anything above 0.5 is labeled as positive. Table 10.6 displays the contingency matrix based on the results of the logistic regression sentiment analysis. The logistic regression model results in the same accuracy as the NB and SVM sentiment analysis models, or 65%.

To understand the results, we take a closer look at some of the accurate and inaccurate predictions based on the testing data. As shown in Fig. 10.9, the logistic regression sentiment analysis model also has difficulty classifying ambiguous reviews but performs as expected with clearer sentiments. This model, like the NB

Table 10.6 Logistic regression contingency matrix classification analysis

		Actual		
		-	+	Total
Predicted	-	6	3	9
	+	4	7	11
	Total	10	10	20

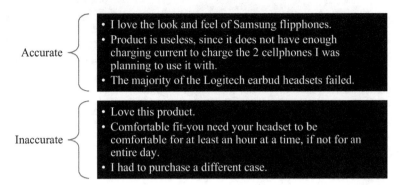

Accurate
- I love the look and feel of Samsung flipphones.
- Product is useless, since it does not have enough charging current to charge the 2 cellphones I was planning to use it with.
- The majority of the Logitech earbud headsets failed.

Inaccurate
- Love this product.
- Comfortable fit-you need your headset to be comfortable for at least an hour at a time, if not for an entire day.
- I had to purchase a different case.

Fig. 10.9 Examples of accurate and inaccurate predictions using logistic regression

model, also struggles to accurately identify phrases such as, "Love this product," which appears to have a clearly positive sentiment.

10.3 Sentiment Analysis Performance: Considerations and Evaluation

As the examples using the learning approach illustrate, there are some difficulties with this method. The consistent accurate and inaccurate predictions, both positive and negative, are displayed in Table 10.7. These three models are all better able to accurately predict positive sentiments than negative sentiments. Learning models perform best when they are trained using a large amount of data. For instance, the NB and logistic regression models both fail to accurately identify "Love this product" as a positive review. In taking a closer look at the DTM, we find that the positive term in the sentence, *love*, appears only once in only one document. The prevalence of the term is clearly insufficient for accurate classification using these two sentiment analysis model methods.

With straightforward sentences, sentiment analysis appears simple. Unfortunately, such sentences may be rare. People use language in many ways that, at times, makes it hard for even human readers to agree on the sentiment (Mullich 2013).

For example, take the following restaurant review:

Their bread is delicious, it's nicely buttered and warm with a light browned top-split. However, the lobster itself wasn't impressive. Don't get me wrong, it's still yummy and I liked that the lobster was not drenched in butter or mayonnaise, but I feel like I remember the bread more than the lobster meat. Their lobster meat here tasted fresh, but wasn't that sweet and overall was a little bland to me. I also tried their Lobster Bisque (Cup $7.50), which I liked. It's pretty creamy and heavy though, in my opinion. Probably more cream in it than there should be.[5]

---

[5]T., Debra. "Maine-Ly Sandwiches." Yelp, Yelp, 2 Jan. 2018, www.yelp.com/biz/maine-ly-sandwiches-houston-3

Table 10.7 Examples of consistent accurate and inaccurate predictions across learning methods for negative and positive sentiments

	Accurate	Inaccurate
Positive	Great product for the price Display is excellent, and camera is as good as any from that year So far so good Absolutely great Excellent starter wireless headset Best of all is the rotating feature, very helpful And the sound quality is great	Fast service
Negative	The majority of the Logitech earbud headsets failed Poor reliability	This phone tries very hard to do everything but fails at its very ability to be a phone In addition, it feels and looks as if the phone is all lightweight cheap plastic

The review contains elements that are both positive and negative, making it difficult to discern if there is an overall positive or negative sentiment to the review. Furthermore, automated sentiment analysis struggles with similes and metaphors. The phrase "avoid like the plague" is easy for humans to understand, but, unless specifically coded, it will likely be overlooked programmatically.

To evaluate the results of sentiment analysis, we can use measures of external validity. Additionally, we can check them against human-coded sentiment. While 100% agreement is unlikely, an agreement rate of at least 80% is considered good (Mullich 2013). To evaluate a text, first select a random subset of documents. Limit the size to an amount manageable by hand. Next, have people manually read and score each document's polarity. Check the human assessments versus the programmatic assignments to evaluate the agreement, with a benchmarking target of approximately 80%.

> **Key Takeaways**
> - Sentiment analysis, also referred to as opinion mining, is a method for measuring or categorizing the polarity of text documents.
> - Sentiment analysis can be lexicon-based, which relies on preset dictionaries for sentiment identification or learning-based, in which machine learning techniques are employed to identify the sentiment of text documents.
> - General Inquirer, AFINN, OpinionFinder, and SentiWordNet are examples of lexicons used in sentiment analysis.
> - Naïve Bayes, support vector machines, and logistic regression are demonstrated as learning-based approaches to sentiment analysis.

References

Appel, O., Chiclana, F., & Carter, J. (2015). Main concepts, state of the art and future research questions in sentiment analysis. *Acta Polytechnica Hungarica, 12*(3), 87–108.

Baccianella, S., Esuli, A., & Sebastiani, F. (2010, May). Sentiwordnet 3.0: An enhanced lexical resource for sentiment analysis and opinion mining. In *LREC* (Vol. 10, No. 2010, pp. 2200–2204).

Dave, K., Lawrence, S., & Pennock, D. M. (2003, May). Mining the peanut gallery: Opinion extraction and semantic classification of product reviews. In *Proceedings of the 12th International Conference on World Wide Web* (pp. 519–528). ACM.

Dua, D., & Karra Taniskidou, E. (2017). *UCI machine learning repository* [http://archive.ics.uci.edu/ml]. Irvine: University of California, School of Information and Computer Science.

Esuli, A., & Sebastiani, F. (2007). SentiWordNet: A high-coverage lexical resource for opinion mining. *Evaluation*, 1–26.

Feldman, R. (2013). Techniques and applications for sentiment analysis. *Communications of the ACM, 56*(4), 82–89.

Ignatow, G., & Mihalcea, R. (2016). *Text mining: A guidebook for the social sciences*. Los Angeles: Sage Publications.

Kantardzic, M. (2011). *Data mining: concepts, models, methods, and algorithms*. John Wiley & Sons.

Kumar, A., & Sebastian, T. M. (2012). Sentiment analysis: A perspective on its past, present and future. *International Journal of Intelligent Systems and Applications, 4*(10), 1.

Liu, B. (2012). Sentiment analysis and opinion mining. *Synthesis Lectures on Human Language Technologies, 5*(1), 1–167.

Mullich, J. (2013, February 4). Guide to sentiment analysis. Data Informed, data-informed.com/guides/guide-to-sentiment-analysis/.

Nasukawa, T., & Yi, J. (2003, October). Sentiment analysis: Capturing favorability using natural language processing. In *Proceedings of the 2nd International Conference on Knowledge Capture* (pp. 70–77). ACM.

Nielsen, F. Å. (2011). A new ANEW: Evaluation of a word list for sentiment analysis in microblogs. arXiv preprint arXiv:1103.2903.

Silge, J., & Robinson, D. (2016). Tidytext: Text mining and analysis using tidy data principles in R. *JOSS, 1*(3), 37.

Silge, J., & Robinson, D. (2017). *Text mining with R: A tidy approach*. Sebastopol: O'Reilly Media, Inc.

Stone, P., Dunphry, D., Smith, M., & Ogilvie, D. (1966). *The general inquirer: A computer approach to content analysis*. Cambridge, MA: MIT Press.

Wilson, T., Wiebe, J., & Hoffmann, P. (2005, October). Recognizing contextual polarity in phrase-level sentiment analysis. In *Proceedings of the Conference on Human Language Technology and Empirical Methods in Natural Language Processing* (pp. 347–354). Association for Computational Linguistics.

Further Reading

For an example of implementing lexicon-based sentiment analysis in R, see Chap. 13. For an example of implementing learning-based sentiment analysis in RapidMiner, see Chap. 15.

Part IV
Communicating the Results

Chapter 11
Storytelling Using Text Data

Abstract This chapter explores the concept of data storytelling, an approach used to communicate insights to an audience to inform, influence, and spur action. A storytelling framework is included for reference and can be used to develop, focus, and deliver the most important concepts from an analysis that should be conveyed within a narrative.

Keywords Storytelling · Insights · Narrative · Reporting

11.1 Introduction

Storytelling has been embedded in the human experience since the beginning of time. The influential art of the spoken word dates back thousands of years—an oral tradition breathing life into information through an entertaining or compelling narrative. Stories promote social interaction and connection, affirm perceptions and beliefs, and allow people to make sense of complex concepts. Stories are tools that have the ability to take listeners on an impactful journey. Effective storytelling can turn the mundane into exciting. It has the ability to reimagine the minutiae as the pivotal.

From broadcast journalists, directors, and entertainers to authors, playwrights, and artists, storytelling has always played an important role in society. In fact, some of the world's most iconic speeches have incorporated narratives that captured and inspired audiences. Whether it is William Shakespeare's "to be or not to be" or Abraham Lincoln's "of the people, by the people, for the people," centuries-old words from famous stories still resonate with us today. In more recent years, Pakistani activist Malala Yousafzai shared stories about equality and the rights of women and children, rallying leaders to provide worldwide access to education at the United Nations Youth Assembly. Airbnb—an online platform connecting producers and consumers of accommodations—launched an initiative to leverage storytelling to reach audiences. The organization's new stories feature on its website is a page dedicated solely to customer narratives about Airbnb's services. Packaging information into a story allows concepts to be shared, heard, and felt. Perhaps even more impactfully, when rooted in data, those stories can emerge into something even more transformative.

11.2   Telling Stories About the Data

While storytelling itself dates back to the origins of speech, the concept of data storytelling is a skill that many have yet to conquer. Few talents are as impactful as the ability to tell an engaging story. While the term "data storytelling" has often been used to describe data visualization, as in business intelligence, it extends beyond the use of dashboards, heat maps, and infographics. The objective of data storytelling is to communicate insights to inform, influence, and elicit action. Ultimately, the success of a text analytics application hinges on the ability to communicate the results.

The ability to tell a story with numbers is a highly sought-after skill with demand that is expected to increase in the future. Google's Chief Economist Dr. Hal R. Varian said, "The ability to take data – to be able to understand it, to process it, to extract value from it, to visualize it, to communicate it – that is going to be a hugely important skill in the next decades" (Dykes 2016).

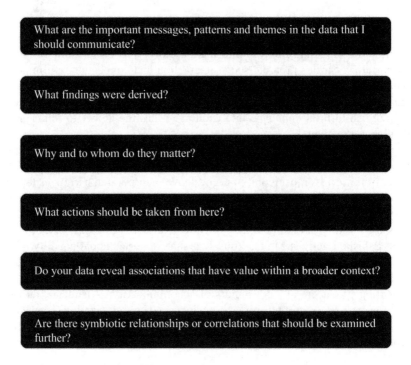

What are the important messages, patterns and themes in the data that I should communicate?

What findings were derived?

Why and to whom do they matter?

What actions should be taken from here?

Do your data reveal associations that have value within a broader context?

Are there symbiotic relationships or correlations that should be examined further?

Fig. 11.1 Questions to ask to identify the key components of the analysis

Developing an effective narrative requires merging data and meaning to convey analysis-driven findings. The data on their own are not enough. As noted, author and data visualization expert Stephen Few said, "Numbers have an important story to tell. They rely on you to give them a clear and convincing voice" (Dykes 2016). Without a story, analyses fall flat. According to Jennifer Aaker, a marketing professor at the Stanford Graduate School of Business, people remember information when it is woven into narratives "up to 22 times more than facts alone" (Aaker 2012).

In business, stories provide audiences with a journey of events that highlights the information they need to know. Data storytelling illuminates the key components of the analysis—the ones most worth sharing. As Fig. 11.1 shows, there are seven important questions to guide the storytelling strategy. Answering these questions will help the analyst package and sequence the findings in a clear way that is useful and of interest to the audience.

Perhaps the most critical aspect of telling a story is knowing the audience. As Chap. 12 indicates, it is important to consider the stakeholders in the analysis. Additionally, the consideration of their interests and priorities can be crucial. Given their roles within or across an organization, how will the message be received and understood? Which audience members can and will influence each other when it comes to making decisions?

When communicating the story, it is also important to be free of bias. Bias occurs when one outcome is subjectively favored. Of course, opinions, explanations, and context can be interwoven with factual information, but including bias when communicating the results of a data-driven analysis can invalidate or weaken a strong, fact-based story.

Data storytelling is critical not only for the audience to retain the key message but also for understanding it. In addition to creating a memorable presentation, it is important to keep the message simple. Philanthropists Bill and Melinda Gates have been acknowledged for their ability to use simple language to make complex concepts understandable. As part of an announcement about a "Breakthrough Energy Coalition" of private investors on a mission to advance clean energy initiatives, Bill Gates detailed the science of climate change in a paper using words that were "remarkably short, simple and free of jargon," explaining the problem and solution in a way that was readily understood at a high school level. Gates is passionate about "learning and communicating what he's learned to galvanize people to action. And [he] cannot persuade if his audiences don't understand him" (Gallo 2015).

While the mastery of skills is crucial to secure a job as a text analyst, the ability to communicate data-driven results effectively to a nontechnical audience and make the findings accessible means that the insights can be shared with a large number of people. A good storyteller eliminates jargon and overly complicated or confusing content, instead communicating only the necessary information that supports and strengthens the central argument. If more detail from the analysis is necessary to answer questions from analytical listeners, including more granular information in an appendix within the presentation or report can be a talking point for post-presentation conversation.

Without communicating the story, valuable opportunities to move the organization forward, secure potential investment or resources for a new strategic direction, and influence decision-making among leaders and colleagues will be lost. A compelling story ignites meaningful dialogue about performance, customers, strengths, and areas where improvement is needed. The result is a well-informed team with an in-depth understanding of organizational performance and clear goals for future direction. Stories share realities but also spur growth, reflection, and further discussion within an organization.

11.3   Framing the Story

11.3.1   Storytelling Framework

Stories are not about sharing statistics or logic. The power of a story lies in its ability to elicit human responses. To hold the attention of any audience, the story needs to be properly framed to engage listeners.

Using a framework to develop and deliver the data story helps the analyst focus on the most important concepts to convey. The following steps, displayed in Fig. 11.2, can guide the strategic process:

1. *Introduce the story.* In this initial step, give the audience a high-level introduction to the analyst and analysis. In other words, describe the end goal. Perhaps the objective is to convey customer sentiment about a specific product or brand. Let the audience know what will be shared, setting the scene for what they can expect.
2. *Connect with the audience.* Engage with the listeners. Who are the people in the room? What is their relationship with one another? Why does this information matter to them? Tailor the delivery to the audience, reminding them that the discussion itself can help the organization progress and achieve its goals.
3. *Deliver the facts.* What are the key data points that should be shared?
4. *Highlight the central argument.* What is the key challenge or component being shared? What does the information mean, and where is the organization now in comparison to where it should be?
5. *Communicate the insight.* Integrate the problem within the context of the organization. What is the relationship between the data and the key challenges? Are there additional meanings within the data that are not explicitly stated?
6. *Identify what is at stake.* If the current situation does not improve, what will the potential outcome be? Who and how many will this result affect?
7. *Share the potential impact.* Illustrate what could be, sharing potential future results. Describe the problem as an opportunity. What does this situation make possible? What is the value proposition or return on investment?
8. *Envision a roadmap to success.* Now that we have the information, how do we use it to get to where we want to go? What needs to be achieved, when, and by whom?

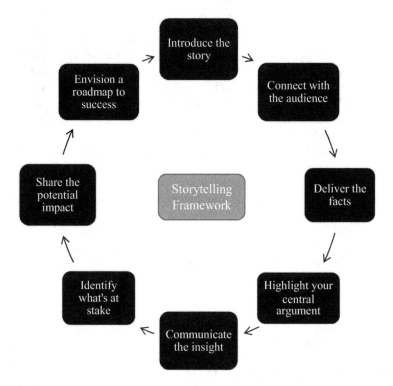

Fig. 11.2 Storytelling framework

11.3.2 Applying the Framework

We will provide an example using the hospitality industry. An extended-stay hotel organization is interested in learning more about the overall journey of its customers—from initial room booking to final checkout. The central question is: How do customers perceive the organization? What is the brand identity? What do customers think about the services offered? Where are the strengths and weaknesses of the overall customer journey?

Data from the organization's registration system, social media platforms, and other sources were compiled to complete a text analytics assessment of customer sentiment with the objective of extracting insights and themes about customers' perceptions. The data show that customers want a more modernized experience and more personalized interactions with the hotel staff. Moreover, the online booking system used to make reservations is difficult to navigate and could benefit from being more user-friendly.

Using the framework, we revisit each step to communicate these insights through data storytelling. Each component of the story framework below includes an example of how the extended-stay hotel might address its data analysis, as in the latent semantic analysis (LSA) example using Python presented in Chap. 14.

1. *Introduce the story.*

   As an organization, we're customer-centric and always aiming to exceed expectations. From the data we've collected and analyzed, we've extracted insights about the overall customer journey—from pre-arrival and check-in to the duration of the stay and checkout—that we'd like to share. These insights provide a heightened awareness of our customers' experiences and an opportunity to understand how our organization and performance are perceived.

2. *Connect with the audience.*

   The data deal with many areas and functions and can improve customer service and marketing efforts and personalize the experience. With valuable insights across various departments, the results can help the organization optimize the effectiveness of its operations. The data can improve our meaningful engagement with our customers.

3. *Deliver the facts.*

   From our analysis of customer sentiment, 67% of our customers are dissatisfied with the interior appearance of the rooms. Words such as *dark*, *outdated*, and *dingy* were mentioned quite frequently.

4. *Highlight your central argument.*

   We need to modernize and enhance the guest experience in order to meet the expectations of today's travelers and expand our market share. The negative sentiments appeared most frequently in comments from people ages 28–44. By listening to the voice of our potential customers within this generation, we can address the problems that emerged from the analysis.

5. *Communicate the insight.*

   We have experienced success in four of our suburban locations, delivering quality experiences to primarily couples and families. However, three of our urban locations that typically serve business travelers and individual guests are underperforming. Those locations are losing customers and spending money on strategies that are not aligned with their target audience. The organization's investments are not targeted toward the right population of potential customers.

6. *Identify what's at stake.*

   If we don't take action on these issues, we will lose market share and damage our reputation, leading to the potential closure of our urban locations and a loss of revenue.

7. *Share the potential impact.*

   Modernizing the guest experience, both technologically and aesthetically, and creating active, social spaces will help us offer people-centric experiences to satisfy the 28–44 age group. By attracting customers within this target segment, we have the opportunity to gain market share, leading to organizational growth, enhanced reputation, and increased revenue.

8. *Envision a roadmap to success.*

We can enhance the experience of our customers by providing a simple, intuitive online booking system and a modern, inviting atmosphere where guests can recharge and socialize. We can attract younger generations of customers by offering a contemporary design and amenities, and a strengthened sense of community, eventually leading to increased market share and an elevated brand.

11.4   Organizations as Storytellers

How do organizations reach their customers through storytelling? Several companies have demonstrated innovative approaches to addressing their data through narratives.

11.4.1   United Parcel Service

UPS' story of innovation is a highlight in the analytics field. The decade-long operation research project—On-Road Integrated Optimization and Navigation (ORION)—ensures that drivers use the most effective delivery routes, based on components such as time, fuel, and distance. When UPS shares the ORION story, it does not describe the program's advanced algorithms and fleet telematics. The story is a structured approach introducing the project's origin, the challenge that was tackled, the methodology, the outcome of the effort, and the overall results.

It was UPS' founder Jim Casey's belief that companies should continually identify ways to improve. That belief, accompanied by the organization's commitment to invest in technology, led to the inception of ORION and the initiative to optimize nearly 55,000 routes in North America. As a result of ORION, UPS drives approximately 100 million fewer miles per year, consumes 10 million gallons less fuel, and reduces its carbon dioxide emissions by 100,000 metric tons. According to the organization's assessment, UPS can save up to $50 million per year by reducing 1 mile per driver per day.

The project not only helped the organization but also was beneficial for its customers, who now have access to personalized services for tailoring home delivery preferences, rerouting shipments, and modifying deliveries. While the undertaking of the project required a complex, technical solution that was far from simple, the organization shares its efforts and results in a way that is easily understood by audiences and customers (UPS 2016).

11.4.2   Zillow

People at Zillow are accomplished storytellers. The online real estate marketplace considers storytelling to be a critical component of connecting with current and future residents. Stephanie Reid-Simons, director of content at Zillow Group, knows that reaching her audience with quality content requires a powerful story and that storytelling is directly associated with return on investment. "Storytelling in any business is vital, but it is especially so in the apartment industry. You are creating homes, lifestyles and a rich community with neighborhood players. Your residents and potential residents are hungry for information and ideas about your area and their interests."

To reach potential customers in the apartment industry, Zillow uses storytelling to trigger emotional responses from audiences. Stories about creating a sense of community and a place to call home resonate with hopeful residents who are dreaming of the perfect neighborhood and the opportunity to develop a lifestyle dedicated to the people and things most important to them. Zillow's strategic approach to storytelling uses an emotional hook to connect and engage with more people, expanding the organization's marketing reach over time Schober (2016).

11.5 Data Storytelling Checklist

The analyst can use this checklist to guide his or her storytelling efforts.
Introduce the story

- Why are you there?
- What information do you intend to share?

Connect with the audience

- Why does this information matter to them?
- What is the outcome of the discussion?

Deliver the facts

- What are the key data points that should be shared?

Highlight your central argument

- What is the key challenge or component that needs to be changed? What does your argument mean in terms of the organization and where are things now in comparison to where they should be?

Communicate the insight

- What is the relationship between the data and the key challenges?
- Are there additional meanings within your data that are not explicitly stated?

Identify what is at stake

- If changes or improvements are not made, what is the potential outcome?
- Who and how many stakeholders will be affected by this result?

Share the potential impact

- What does this situation make possible?
- What is the value proposition or return on investment?

Envision a roadmap to success

- What needs to be achieved?
- How do we get there?

Key Takeaways
- Data storytelling involves communicating insights from the data to inform, influence, and spur action.
- A critical aspect of storytelling is knowing the audience and stakeholders—and what matters most to them.
- A good storyteller eliminates technical jargon and overwhelming content, communicating only the necessary information that supports and strengthens the central argument.
- To hold the attention of the audience and elicit action, a story should be properly framed in a way that engages listeners. Using a framework to develop and deliver the story will help the analyst focus on the most important concepts he or she wants to convey.

Acknowledgements The authors thank Diana Jones, Associate Director of the Business Analytics Solutions Center and the Dornsife Office for Experiential Learning, at the LeBow College of Business, Drexel University for contributing this chapter to the book.

References

Aaker, J. (2012). *How to harness stories in business*. Stanford Business School. Retrieved May 29, 2018, from https://www.gsb.stanford.edu/faculty-research/case-studies/how-harness-stories-business

Dykes, B. (2016, March 31). Data storytelling: The essential data science skill everyone needs, Forbes. Retrieved May 29, 2018, from https://www.forbes.com/sites/brentdykes/2016/03/31/data-storytelling-the-essential-data-science-skill-everyone-needs/2/#3ea7514b7ee0

Gallo, C. (2015, December 11). Bill and Melinda Gates brilliantly explain complex stuff in simple words. Retrieved May 29, 2018, from https://www.forbes.com/sites/carminegallo/2015/12/11/bill-and-melinda-gates-brilliantly-explain-complex-stuff-in-simple-words/#7d436bde30b8

Schober, L. (2016, April). Zillow Group. Retrieved May 29, 2018, from https://www.zillow.com/multifamily-knowledge-center/tips-trends-training/apartment-marketing/whats-the-roi-of-storytelling/

UPS. (2016). Retrieved May 29, 2018, from https://www.pressroom.ups.com/pressroom/ContentDetailsViewer.page?ConceptType=Factsheets&id=1426321616277-282

Further Reading

Gabriel, Y. (2000). *Storytelling in organizations: Facts, fictions, and fantasies*. Oxford: Oxford University Press.

Gargiulo, T. L. (2005). *The strategic use of stories in organizational communication and learning*. Armonk: M.E Sharpe.

Chapter 12
Visualizing Analysis Results

Abstract Text visualizations are the topic for this chapter. The chapter begins with general techniques to help create effective visualizations. From there, it moves to common visualizations used in text analysis. The chapter describes heat maps, word clouds, top term plots, cluster visualizations, topics over time, and network graphs.

Keywords Text analytics · Text visualization · Word clouds · Document visualization · Document networks · Text clouds

With the analysis complete and the storytelling strategy in place, it is time to share the results. Visualizing the results of the analysis is an integral part of the storytelling strategy, as covered in Chap. 11. According to Kucher and Kerren (2015, p. 476), text visualization includes techniques for visually representing raw text data or the results of text analysis methodologies. Effective and meaningful visualizations convince audiences of the message and help them understand the findings. In this chapter, we will discuss the elements that make an effective visualization and the factors that should be considered when choosing a visual representation of the results of the text analysis.

Powerful visualizations help the audience understand the presenter's message. Comprehending the patterns and trends is easier when looking at a visualization than when viewing just the raw data ("Data Visualization: What it is and Why it Matters" 2017). The importance of visualization in data analysis spurred the creation of a subfield of analytics in the early 2000s known as visual analytics (Ellis and Mansmann 2010). The use of visuals can play an important role in communicating the results of the text data analysis to the audience, reduce their complexity, and increase the cohesion and clarity of the findings (Keim et al. 2006).

In this chapter, we cover the fundamentals of effective visualization. In doing so, the chapter discusses the importance of text visualization strategies. Then we present examples of text visualizations at the word, document, category/theme, and corpus levels.

M. Anandarajan et al., *Practical Text Analytics*, Advances in Analytics and Data Science 2, https://doi.org/10.1007/978-3-319-95663-3_12

12.1   Strategies for Effective Visualization

We begin with several important considerations when trying to create an impactful visualization. As Fig. 12.1 shows, while constructing the visualization, the presenter should be purposeful, know the audience, solidify the message, plan and outline, keep it simple, and try to focus the attention of the viewer.

12.1.1   Be Purposeful

Visualizations are typically grouped into either explanatory or exploratory. Explanatory visualizations explain a topic to the audience and are used for presentations. Exploratory visualizations help the analyst understand the data as he or she works through the analysis (Knaflic 2015a, b, c). Given that we are discussing prepping for a presentation, we focus on explanatory visualizations. The goal of the visualization is to prove a belief. Therefore, it centers on the important points and does not show every factor or step in the analysis (Knaflic 2015a, b, c). In other words, the visualization should be purposeful. Rather than providing the audience with a data dump, the presenter should create a story and walk the audience through the proof. The visualization is an aid for telling a story and supporting the points the presenter wants to make (Knaflic 2017).

12.1.2   Know the Audience

Next, the presenter should consider the target audience of the analysis and construct the visualization with them in mind. By understanding the audience, the presenter can tailor the visualization to maximize the chances that they will engage with the story. The purpose of the visualization is to convince them. To do that, the visualization must fit their knowledge base and goals for the project. If the presenter understands their goals, he or she can tailor the visualization to provide the correct information in the correct form (Knaflic 2015a, b, c).

12.1.3   Solidify the Message

With an understanding of the purpose and the audience in hand, the presenter should now create the "big idea" (Knaflic 2015a, b, c). In other words, he or she must identify the main points to tell the audience (Knaflic 2015a, b, c). To accomplish this goal, the presenter should ask himself or herself detailed questions. This approach is an effective way to narrow the focus to the most important and impactful items to emphasize. For example, the presenter might begin with a consideration of the goal

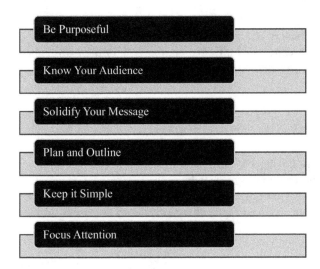

Fig. 12.1 Strategies for text analytics visualizations

of the project. A broad start limits the responses to the essential parts of the project. The presenter can then narrow these parts down to the most impactful elements of the analysis (Berinato 2016).

12.1.4 Plan and Outline

The next step is to spend time planning and outlining the appearance of the visualization. While the final result may differ, the goal is to have a set of notes or sketches on which to focus. Taking this step will help ensure that the presenter is not distracted by software options. It is the notes that should guide the presenter, not the recommended graph in Excel (Knaflic 2015a, b, c).

The notes should include sketches of the presentation. This visualization should make it easy for others to understand the presenter's ideas (Berinato 2016). If one train of thought does not make sense, the presenter should start over. With each new iteration, the goal is to move closer to a concise presentation of the message.

12.1.5 Keep It Simple

Simple visualizations offer the presenter the best chance of helping the audience understand the desired message. Audiences do not need or want to commit a great deal of effort to understanding complex visualizations or ones that they regard as complex. When faced with the need to do so, they are likely to ignore the visualization. The presenter's goal should be to minimize the effort the audience needs to expend on understanding the visualization (Knaflic 2015a, b, c). Therefore, the presenter should include only the items that increase the understanding of the message. By leaving out extraneous information, the visualization will entice the audience, not repel them (Knaflic 2015a, b, c).

12.1.6   Focus Attention

The next way for the presenter to maximize the chances that the audience will understand his or her message is to focus their attention with pre-attentive attributes. Pre-attentive attributes highlight specific locations and draw the audience's focus to that spot (Knaflic 2015a, b, c). In this context, they are visual attributes such as size, color, and position. When used judiciously, the audience's mind is unconsciously drawn to this highlighted location.

12.2   Visualization Techniques in Text Analytics

The high dimensionality of text data, including large document collections with many words, can create some difficulty with respect to the visualization of analysis results. Yang et al. (2008) report that some common real-world, text-focused visualizations include histograms, co-occurrence matrices, cluster maps, and network graphs.

The granularity of the visualization may vary depending on the purpose of the analysis. However, in most cases the presenter should begin with the overall picture before focusing on the more detailed aspects of the analysis (Gan et al. 2014). In this section, we will revisit many of the plots that have been presented in this book and introduce some additional plots that can be used to help the presenter convey the intended message and tell the story of the analysis. We begin by visualizing the big picture at the document collection level and then narrow our focus to the category/ theme level and document level.

12.2.1   Corpus/Document Collection-Level Visualizations

Corpus-level representations can include heat maps, document relationship networks, and document similarity maps (Gan et al. 2014). Document maps are a useful method to present a spatial representation of document similarity (Alencar et al. 2012). First, we consider visualizations at the highest granularity level: the full corpus. Regardless of the analysis method used, the goal is to illustrate the big picture: all of the documents or words. In our small-scale example in Chap. 5 with 10

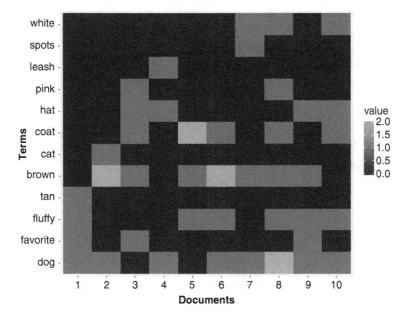

Fig. 12.2 Heat map visualization

documents and 12 terms, we used a heat map displaying the frequency of terms in the documents, which is displayed again in Fig. 12.2. A heat map can be used, as opposed to presenting a matrix with frequency values. If applicable, the color scheme can be altered to align with the focus of the analysis or to match corporate branding protocols. Heat maps are useful in small-scale analysis but are less informative or effective for larger-scale analysis.

Another type of plot that can be used, which can represent term frequencies at the corpus level, is a word cloud. Word clouds are a common visualization created for text analyses. Usually, they display the most frequent words in the corpus. Each word is sized by its frequency. The words are often displayed with colors or fonts that make word clouds superficially attractive and fun (Heimerl et al. 2014). They provide audiences with a summary of word usage, but they are limited beyond this basic overview (Heimerl et al. 2014). They do not include any information about the interrelationship between words or any exact measurement information.

Word clouds have many applications in text analytics, because they are visually appealing and a simple alternative to presenting a large, sparse DTM to the audience. There are many elements of word clouds such as colors and shapes that can be customized for the analysis. For example, in the document collection used in Chaps. 8 and 9, in which dog owners describe their dog's physical appearance, the intended audience is probably fellow dog lovers. For this reason, as Fig. 12.3 illustrates, we create a word cloud in the shape of a dog's paw to appeal to our audience. The word cloud gives us an idea of the global term frequency, without presenting it in a more formal, standard barplot.

Fig. 12.3 Word cloud of dog descriptions in the shape of a paw

12.2.2   Theme and Category-Level Visualizations

The next level of granularity is at the category or theme level. At this level, we want to consider LSA dimensions, clusters, topics, categories, and sentiments.

12.2.2.1   LSA Dimensions

The output of an LSA is a multidimensional space created by using SVD for dimension reduction. The LSA space has many dimensions and may be difficult to conceptualize or understand for people who are unfamiliar with it. Rather than focus on the big picture, we can visualize the terms that feature most in LSA dimensions. Figure 12.4 displays the top 10 words in each of the four LSA dimensions in the analysis using Python in Chap. 15. From the plot, we have an at-a-glance view of all four dimensions from which we can try to characterize the latent information in each dimension further.

12.2.2.2   Cluster-Level Visualizations

In analyses that cluster documents using HCA, the dendrogram resulting from the analysis can give the audience a clear and concise idea of naturally occurring groupings of documents in the document collection. As covered in Chap. 7, the

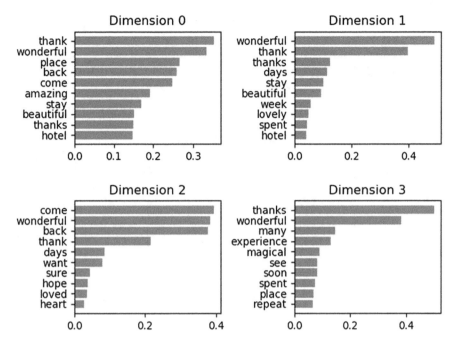

Fig. 12.4 Plots of top terms in first four LSA dimensions

dendrogram is a benefit of HCA analysis and can be informative in the analysis of both word and document clusters. However, for large-scale analysis, it can be confusing and difficult for the audience to interpret. When choosing the visualization, the presenter should keep in mind both the audience and the size of the data to determine the most effective way to relay the findings. Figure 12.5 contains two versions of the same dendrogram based on the seven-cluster HCA solution found in the chapter. In both versions, we use color to distinguish the clusters and focus the reader's attention. In the first dendrogram, we alter only the colors of the rectangles distinguishing the clusters. In the second dendrogram, we expand the use of the seven different colors and alter the orientation of the plot.

12.2.2.3 Topic-Level Visualizations

At the topic level, plots of the most important terms in a topic can have a great deal of explanatory power. In Fig. 12.6, we use different colors to differentiate the various topics and plot the most important topic words. The top three terms in Topic 1 are *long*, *tall*, and *short*, indicating that they have to do with the physical description of the dog's height. Topic 2 is dominated by *white*, and we would expect to find this topic discussed in most of the Bichon Frise documents. *Weight* and *pound* are the top two terms in Topic 3, indicating that this topic deals with the size of the dog, in terms of weight. Finally, Topic 4 appears to describe the physical attributes of the dogs.

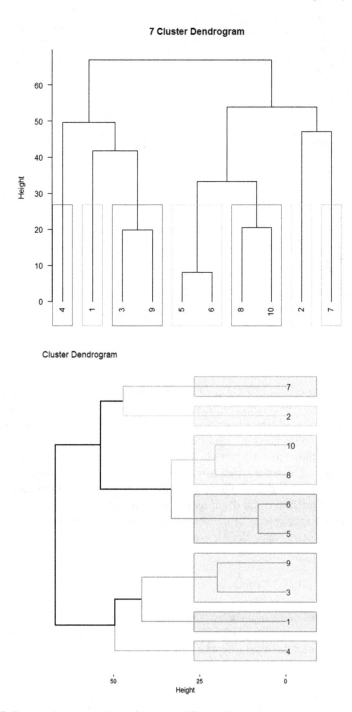

Fig. 12.5 Two dendrogram versions of the same HCA solution

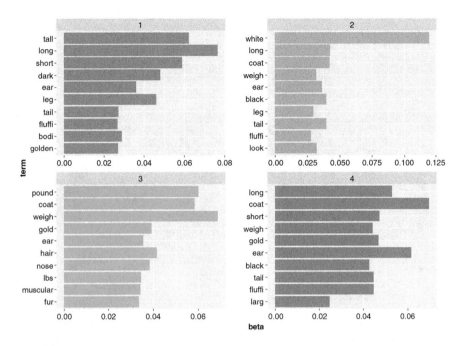

Fig. 12.6 Top ten terms per topic, four-topic model

Visualizations are an important way to not only present the results of the analysis but also let the audience see the trends and patterns in the data and analysis for themselves. Figure 12.7 depicts the temporal topic trends based on an STM analysis of complaints about vehicle defects for a topic describing sudden acceleration. By looking at the expected topic proportions over time, we can visualize trends over time to put our findings into context. This visual aid offers a concise way to contextualize the topic analysis.

12.2.2.4   Category or Class-Level Visualizations

As shown in Fig. 12.8, document network plots can help us spatially understand connections among documents that might otherwise be missed. In the top figure, we build an adjacency-based network plot based on the similarity or closeness of the documents in the document collection presented in Chap. 8. We use the default plot settings. From this visualization of the document network, we can see that there are some documents that clump together, suggesting there are groups of similar documents. However, the plot does not inform us about which documents might be similar.

In the bottom plot, we incorporate color-coded class information. Now, we have a much clearer view of the document collection. As we can see, the documents describing Bichon Frises are very similar. On the other hand, the Golden Retriever

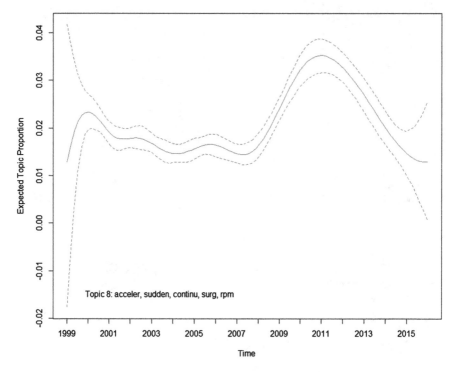

Fig. 12.7 Plot of expected topic proportions over time

document network is more spread out. The Dachshund and Great Dane documents do not display clearly defined groupings, suggesting that we would expect there to be similarities in these documents. In addition to adding the color schema and legend, we also muted the impact of the edge lines connecting the document nodes by changing the black lines to a light gray. This color change draws the audience's focus forward to the document nodes.

12.2.2.5   Sentiment-Level Visualizations

Sentiment analysis aims to classify or measure the polarity of documents in a document collection. Color schemes, such as those in Fig. 12.9, can be chosen along a spectrum of polarity measures, and different colors, such as red and green, can be utilized to separate the sentiments. This figure provides separate word clouds for the positive and negative sentiments, with positive in green and negative in red. If the neutral sentiment were also included, it could be plotted in amber to follow the stoplight theme. Displaying the word clouds side by side lets the viewer see the big picture of the analysis.

As we have seen, the use of word clouds can be as generic or customized as the presenter chooses. For instance, we can create a word cloud for online reviews

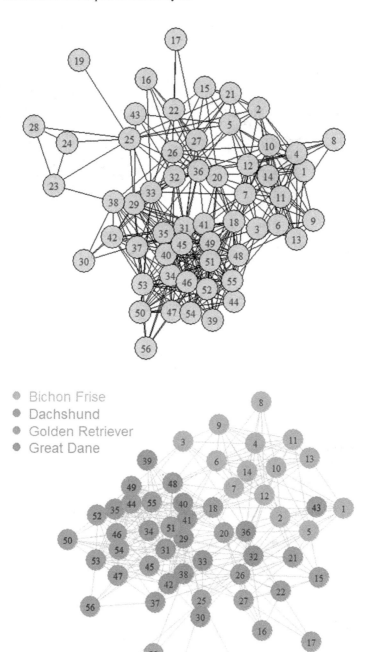

Fig. 12.8 Two versions of the same document network plot

Fig. 12.9 Word clouds of positive and negative words in review sample

Fig. 12.10 Five-star word cloud of reviews

receiving five stars. For online reviews, the most positive reviews have five stars, and the most negative reviews have one star. We can use this domain knowledge to provide an alternative representation of the terms in positive reviews, as shown in Fig. 12.10.

12.2.3   Document-Level Visualizations

Document-level visualizations, which focus on words and phrases in a document, can be represented by tag clouds, word clouds, or enhanced text (Gan et al. 2014). To see their power in action, look at the following reviews.[1] First, we consider the reviews as normal text:

---

[1] www.yelp.com/biz/summer-china-diner-houston-4

- *Ordering took a while*. When the waiter came over, he said they were really busy. There were only four other tables occupied. We both ordered fried rice and one of us was served white rice.
- The soft spring rolls were 3″ or 4″ long and not very good. I ordered the beef and broccoli, and the beef was literally the worst I have ever eaten. *Mediocre service as well*. Sweet tea is good.
- The food was very tasty, hot, and fresh. *The waiter was not attentive*. The food was freshly cooked and came out in 15 min piping hot.

When all of the text is the same in terms of color and weight, there is no clear message. However, if we bold certain sections of the text, the viewer's focus shifts to those words.

- *Ordering took a while.* When the waiter came over, he said they were really busy. There were only four other tables occupied. We both ordered fried rice and one of us was served white rice.
- The soft spring rolls were 3″ or 4″ long and not very good. I ordered the beef and broccoli, and the beef was literally the worst I have ever eaten. *Mediocre service as well*. Sweet tea is good.
- The food was very tasty, hot, and fresh. *The waiter was not attentive*. The food was freshly cooked and came out in 15 min piping hot.

Based on this emphasis, it is clear that the presenter is making a statement about the service at this restaurant. Adding color to these sectors can underscore this point even further.

- **Ordering took a while** When the waiter came over, he said they were really busy. There were only four other tables occupied. We both ordered fried rice and one of us was served white rice.
- The soft spring rolls were 3″ or 4″ long and not very good. I ordered the beef and broccoli, and the beef was literally the worst I have ever eaten. ***Mediocre service as well***. Sweet tea is good.
- The food was very tasty, hot, and fresh. ***The waiter was not attentive***. The food was freshly cooked and came out in 15 min piping hot.

By changing only an attribute or two, we make it easy for the audience to digest the message (Knaflic 2015a, b, c).

Key Takeaways
- Strategies for effective visualizations include be purposeful, know your audience, solidify your message, plan and outline, keep it simple, and focus attention.
- Visualizations at different levels of detail should be provided to convey the important high-level and low-level information to your audience.

References

Alencar, A. B., de Oliveira, M. C. F., & Paulovich, F. V. (2012). Seeing beyond reading: A survey on visual text analytics. *Wiley Interdisciplinary Reviews: Data Mining and Knowledge Discovery, 2*(6), 476–492.

Berinato, S. (2016). *Good charts: The HBR guide to making smarter, more persuasive data visualizations*. Cambridge, MA: Harvard Business Review Press.

Data visualization: What it is and why it matters. SAS, SAS Institute, Inc. (2017, December 1). www.sas.com/en_us/insights/big-data/data-visualization.html

Ellis, G., & Mansmann, F. (2010). Mastering the information age solving problems with visual analytics. In *Eurographics* (Vol. 2, p. 5).

Gan, Q., Zhu, M., Li, M., Liang, T., Cao, Y., & Zhou, B. (2014). Document visualization: An overview of current research. *Wiley Interdisciplinary Reviews: Computational Statistics, 6*(1), 19–36.

Heimerl, F., Lohmann, S., Lange, S., & Ertl, T. (2014, January). Word cloud explorer: Text analytics based on word clouds. In *System Sciences (HICSS), 2014 47th Hawaii International Conference on* (pp. 1833–1842). IEEE.

Keim, D. A., Mansmann, F., Schneidewind, J., & Ziegler, H. (2006, July). Challenges in visual data analysis. In *Information Visualization. IV 2006. Tenth International Conference on* (pp. 9–16). IEEE.

Knaflic, C. N. (2015a). *Storytelling with data: A data visualization guide for business professionals*. Hoboken: John Wiley & Sons.

Knaflic, C. N. (2015b, May 13). Tell your audience what you want them to know. Storytelling with data. Retrieved April 14, 2018, from http://www.storytellingwithdata.com/blog/2015/05/tell-your-audience-what-you-want-them

Knaflic, C. N. (2015c, June 03). Audience, audience, audience. Storytelling with data. Retrieved April 14, 2018, from http://www.storytellingwithdata.com/blog/2015/06/audience-audience-audience

Knaflic, C. N. (2017, September 7). My guiding principles. Storytelling with data. Retrieved April 14, 2018, from http://www.storytellingwithdata.com/blog/2017/8/9/my-guiding-principles

Kucher, K., & Kerren, A. (2015, April). Text visualization techniques: Taxonomy, visual survey, and community insights. In *Visualization Symposium* (PacificVis), 2015 IEEE Pacific (pp. 117–121). IEEE.

Yang, Y., Akers, L., Klose, T., & Yang, C. B. (2008). Text mining and visualization tools–impressions of emerging capabilities. *World Patent Information, 30*(4), 280–293.

Further Reading

For an example of visualizing text analytics results in SAS Visual Text Analytics, see Chap. 16.

Part V
Text Analytics Examples

Chapter 13
Sentiment Analysis of Movie Reviews Using R

Abstract In this chapter, the reader is presented with a step-by-step lexicon-based sentiment analysis using the R open-source software. Using 1,000 movie reviews with sentiment classification labels, the example analysis performs sentiment analysis to assess the predictive accuracy of built-in lexicons in R. Then, a custom stop list is used and accuracy is reevaluated.

Keywords Sentiment analysis · Opinion mining · Online consumer reviews (OCR) · R · RStudio · Open-source

13.1 Introduction to R and RStudio

In this example, we will use the R programming language (R Development Core Team 2008) and its integrated development environment (IDE), RStudio. R is a free statistical computing language with user-contributed libraries. It is available for download at https://www.r-project.org/. RStudio is a free and popular IDE that makes many processes easier. It is available to download at https://www.rstudio.com/. The RStudio workspace is shown in Fig. 13.1.

To follow and replicate the analysis, we present the blocks of code and their output with a light background shading, while the written text has no background.

We will use three packages in R for our analysis: tidytext, dplyr, and ggplot2. The tidytext package will be used to conduct the basic sentiment analysis.[1]

To install and load the necessary packages used in this analysis, run the following lines:

```
install.packages(c('tidytext', 'dplyr', 'ggplot2'))
library(ggplot2) #load the ggplot2 package
library(tidytext) #load the tidytext package
library(dplyr) #load the dplyr package
```

---

[1] For additional information on tidytext and more examples, consult http://tidytextmining.com/

© Springer Nature Switzerland AG 2019   193
M. Anandarajan et al., *Practical Text Analytics*, Advances in Analytics and Data Science 2, https://doi.org/10.1007/978-3-319-95663-3_13

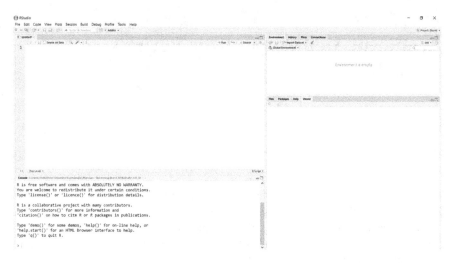

Fig. 13.1 RStudio IDE

13.2 SA Data and Data Import

We use 1,000 Internet Movie Database (IMDB) reviews downloaded from https://archive.ics.uci.edu/ml/machine-learning-databases/00331/. The description of the dataset is available at https://archive.ics.uci.edu/ml/datasets/Sentiment+Labelled+Sentences#.

```
download.file("https://archive.ics.uci.edu/ml/machine-learning-databases/00331
/sentiment%20labelled%20sentences.zip", "sentiment labelled sentences.zip")
trying URL 'https://archive.ics.uci.edu/ml/machine-learning-databases/00331/se
ntiment%20labelled%20sentences.zip'
Content type 'application/zip' length 84188 bytes (82 KB)
downloaded 82 KB

td = tempdir()
unzip("sentiment labelled sentences.zip", files="sentiment labelled sentences/
imdb_labelled.txt", exdir=td, overwrite=TRUE)
fpath = file.path(td, "sentiment labelled sentences/imdb_labelled.txt")
imdb <- read.table(fpath, col.names = c("comment","positive_flag"), quote = ""
, comment.char = "", sep = "\t", stringsAsFactors=FALSE)
imdb$review_id <- 1:nrow(imdb)
```

The additional parameters in the read.table function tell R that there are no quotes (quote = ""), there is no comment character (comment.char = ""), the file is tab delimited (sep = \t), and string variables should not be automatically converted to factor or categorical variables (stringsAsFactors = FALSE). These settings allow R

Fig. 13.2 RStudio workspace with imdb dataframe in the global environment

to read the file properly. We also define the names of the two columns in the dataframe (col.names = c("comment," "positive_flag")). Then, we add an additional column, which includes the unique review ID number.

Once this step is completed, there will be a dataframe in the environment called imdb, which includes 1,000 observations and 3 variables, as depicted in Fig. 13.2. The first column or variable is a series of text comments—the comments to analyze. The second variable is a negative or positive polarity indicator that we can use to judge our model, with 0s indicating negative sentiment and 1 s indicating positive sentiment. The third variable is the ID number.

To get a feel for the data, we can take a closer look, either using str or view. The str function tells us how many documents and variables are in the dataframe, or R object, and gives the first 10 observations for each of the variables. When using the view function, a separate table is opened in the workspace with the data, as shown in Fig. 13.3.

```
str(imdb)
'data.frame':    1000 obs. of  3 variables:
 $ comment       : chr  "A very, very, very slow-moving, aimless movie about a
distressed, drifting young man.  " "Not sure who was more lost - the flat char
acters or the audience, nearly half of whom walked out.  " "Attempting artines
s with black & white and clever camera angles, the movie disappointed - became
even more ridi"| __truncated__ "Very little music or anything to speak of.  "
...
 $ positive_flag: int  0 0 0 0 1 0 0 1 0 1 ...
 $ review_id    : int  1 2 3 4 5 6 7 8 9 10 ...
View(imdb)
```

Fig. 13.3 Table formatted data using the view function

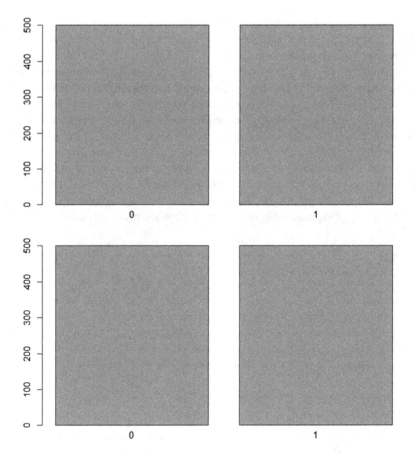

Fig. 13.4 Frequency of negative (0) and positive (1) reviews

Next, let's visualize how many reviews are positive or negative using the barplot function. As the barplot in Fig. 13.4 shows, the dataset is evenly split, with 500 positive and 500 negative reviews. Now that we understand the dataset better, we can start our analysis.

```
barplot(table(imdb$positive_flag))
```

13.3 Objective of the Sentiment Analysis

Defining the purpose of the analysis is our first step. Here, we conduct a data-driven analysis to describe the sentiment analysis process in R. Since the movie reviews include sentiment labels, we can explore which lexicon classifies the reviews best.

The tidytext library comes with three lexicons: NRC, bing, and AFINN. The first lexicon, referred to in tidytext as nrc (Mohammad and Turney 2013), categorizes 13,901 terms as positive or negative and/or by emotion, such as trust, fear, and sadness. We can preview the NRC lexicon using the get_sentiments function and specifying the parameter lexicon equal to nrc.

```
get_sentiments(lexicon="nrc")
# A tibble: 13,901 x 2
          word sentiment
         <chr>     <chr>
 1      abacus     trust
 2     abandon      fear
 3     abandon  negative
 4     abandon   sadness
 5   abandoned     anger
 6   abandoned      fear
 7   abandoned  negative
 8   abandoned   sadness
 9 abandonment     anger
10 abandonment      fear
# ... with 13,891 more rows
```

The bing lexicon (Hu and Liu 2004) includes 6,788 words, which are classified as either positive or negative. We can preview the bing lexicon using the get_sentiments function and specifying the parameter lexicon equal to bing.

```
get_sentiments(lexicon="bing")
# A tibble: 6,788 x 2
          word sentiment
         <chr>     <chr>
 1     2-faced  negative
 2     2-faces  negative
 3          a+  positive
 4     abnormal  negative
 5      abolish  negative
 6   abominable  negative
 7   abominably  negative
 8    abominate  negative
 9  abomination  negative
10        abort  negative
# ... with 6,778 more rows
```

The AFINN lexicon (Nielsen 2011), which is described in Chap. 4, assigns numerical sentiment scores to 2,476 terms. We can preview the AFINN lexicon using the get_sentiments function and specifying the parameter lexicon equal to afinn.

```
get_sentiments(lexicon="afinn")
# A tibble: 2,476 x 2
          word score
         <chr> <int>
 1     abandon    -2
 2   abandoned    -2
 3    abandons    -2
 4    abducted    -2
 5   abduction    -2
 6  abductions    -2
 7       abhor    -3
 8    abhorred    -3
 9   abhorrent    -3
10      abhors    -3
# ... with 2,466 more rows
```

13.4   Data Preparation and Preprocessing

Returning to the IMDB review data, we will prepare and pre-process the data prior to our analysis.

13.4.1   Tokenize

We will start with some initial preprocessing of our text data. The tidytext package does a great deal of this work automatically with the unnest_tokens() function. Punctuation is removed, and uppercase letters are changed to lowercase. The function then tokenizes the sentences, with the default token being a single word. By changing the token parameter in the function, n-grams are possible, although we will use the default in our analysis.

The code creating our tidytext object says to use the imdb dataframe to unnest the tokens in the comment column into a new column in tidy format called word. By using the str function to view the structure of the tidy_imdb object, we see that the new dataframe has 3 columns and 14,482 rows. Each row is now a word in a review. By using the head function, we view the first 10 rows, corresponding to tokens from the first review.

```
tidy_imdb <- imdb %>% unnest_tokens(word, comment)
str(tidy_imdb)
'data.frame':   14482 obs. of  3 variables:
 $ positive_flag: int  0 0 0 0 0 0 0 0 0 0 ...
 $ review_id    : int  1 1 1 1 1 1 1 1 1 1 ...
 $ word         : chr  "a" "very" "very" "very" ...
head(tidy_imdb, 10)
    positive_flag review_id     word
1               0         1        a
1.1             0         1     very
1.2             0         1     very
1.3             0         1     very
1.4             0         1     slow
1.5             0         1   moving
1.6             0         1  aimless
1.7             0         1    movie
1.8             0         1    about
1.9             0         1        a
```

13.4.2   Remove Stop Words

Next, we consider removing stop words. The tidytext package comes with a generic stop list in the "stop_words" dataset. Let's check to see if we want to remove the stop words on the list. To do so, we will see if any words exist in the stop words dataset and at least one of the lexicon datasets. Sentiments are a built-in dataset in tidytext that includes four lexicons: NRC, bing, AFINN, and Loughran. We use inner_join, which takes two dataframes and creates a new dataframe that includes only the observations where there is a match between the two. Figure 13.5 depicts an inner join, where the resulting dataframe is made up of the overlap between the two dataframes, shown in the middle in darker blue. To limit our comparison to the three lexicons that we will use, we use the OR operator, "|".

Using the length and unique functions, we discover that there are 78 unique stop words that exist on both the stop words list and the NRC, bing, or AFINN lexicon lists. The number of terms classified for each lexicon list is depicted in Fig. 13.6. As shown, since each term on the NRC list can have one or multiple classifications, the largest number of matches is within this list. Based on the large number of sentiment words that are considered stop words, we will not remove stop words in our analysis.

```
stop_words %>% inner_join(sentiments %>% filter(lexicon=="bing" |
lexicon=="nrc" | lexicon=="AFINN"), by="word")
# A tibble: 151 x 5
         word lexicon.x sentiment lexicon.y score
        <chr>     <chr>     <chr>     <chr> <int>
1       allow     SMART      <NA>     AFINN     1
2       alone     SMART      <NA>     AFINN    -2
3  appreciate     SMART  positive      bing    NA
4  appreciate     SMART      <NA>     AFINN     2
5 appropriate     SMART  positive      bing    NA
6   available     SMART  positive      bing    NA
7     awfully     SMART  negative      bing    NA
8        best     SMART  positive      bing    NA
9        best     SMART      <NA>     AFINN     3
10     better     SMART  positive      bing    NA
# ... with 141 more rows
length(unique(stop_lexicon_match$word))
[1] 78
> barplot(table(stop_lexicon_match$lexicon.y))
```

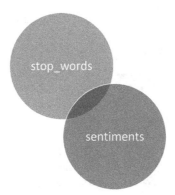

Fig. 13.5 Diagram of an inner join between stop words and sentiments

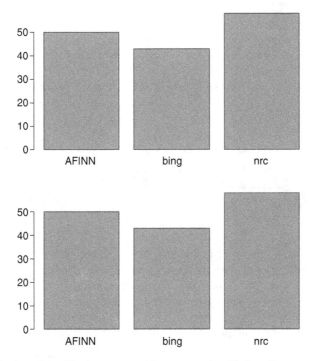

Fig. 13.6 Number of classifications per matching stop word and lexicon lists

13.5 Sentiment Analysis

With the tidytext package, we can match the terms used in our reviews to the lexicon terms using a join to the given lexicon. The polarity, classification, or score for each token is added to the dataframe, which are named according to their lexicon. Again, we use the str function to investigate the size of the dataframes and the variables.

Based on this function, we see that NRC includes additional classes beyond positive and negative, bing provides a positive or negative assessment, and AFINN scores each word for a degree of polarity.

```
nrc_sentiment <- tidy_imdb %>% inner_join(get_sentiments("nrc"), by = "word")
str(nrc_sentiment)
'data.frame':    4432 obs. of  4 variables:
 $ positive_flag: int  0 0 0 0 0 0 0 0 0 0 ...
 $ review_id    : int  1 1 1 1 1 1 1 2 2 2 ...
 $ word         : chr  "aimless" "distressed" "distressed" "young" ...
 $ sentiment    : chr  "negative" "fear" "negative" "anticipation" ...
bing_sentiment <- tidy_imdb %>% inner_join(get_sentiments("bing"),
by = "word")
str(bing_sentiment)
'data.frame':    1696 obs. of  4 variables:
 $ positive_flag: int  0 0 0 0 0 0 0 0 0 1 ...
 $ review_id    : int  1 1 1 2 3 3 3 3 3 5 ...
 $ word         : chr  "slow" "aimless" "distressed" "lost" ...
 $ sentiment    : chr  "negative" "negative" "negative" "negative" ...
afinn_sentiment <- tidy_imdb %>% inner_join(get_sentiments("afinn"),
by = "word")
str(afinn_sentiment)
'data.frame':    1286 obs. of  4 variables:
 $ positive_flag: int  0 0 0 0 0 0 1 0 0 0 ...
 $ review_id    : int  1 2 3 3 3 3 5 6 6 6 ...
 $ word         : chr  "distressed" "lost" "clever" "disappointed" ...
 $ score        : int  -2 -3 2 -2 -3 -2 3 3 -1 -1 ...
```

In our dataset, these are the frequencies of the categories. Beyond classifications as positive and negative, we have *anger, anticipation, disgust, fear, joy, sadness, surprised*, and *trust*. Of the total 704 unique matching terms in the nrc_sentiment dataframe, 622 of those terms are classified as either positive or negative. For this reason, we choose to keep the terms that are classified as either positive or negative by subsetting the nrc_sentiment dataframe.

```
nrc_sentiment <- nrc_sentiment[nrc_sentiment$sentiment %in% c("positive",
"negative"), ]
str(nrc_sentiment)
'data.frame':    1516 obs. of  4 variables:
 $ positive_flag: int  0 0 0 0 0 0 0 0 0 0 ...
 $ review_id    : int  1 1 1 2 3 3 3 3 3 4 ...
 $ word         : chr  "aimless" "distressed" "young" "lost" ...
 $ sentiment    : chr  "negative" "negative" "positive" "negative" ...
```

Next, we need to aggregate the review-level polarity information. To do so, we score each review as a sum of its points. Since AFINN provides sentiment scores, we will total these points. For the bing and NRC lexicons, we count each positive as +1 and each negative as −1. For these two lexicons, we use the ifelse function to accomplish this task and assign the result to a column called "score," as in the AFINN lexicon. The aggregate dataframes include the review numbers and aggregated scores.

```
nrc_sentiment$score <- ifelse(nrc_sentiment$sentiment == "negative", -1, 1)

nrc_aggregate <- nrc_sentiment %>% select(review_id, score) %>%
group_by(review_id) %>% summarise(nrc_score = sum(score))

bing_sentiment$score <- ifelse(bing_sentiment$sentiment == "negative", -1, 1)

bing_aggregate <- bing_sentiment %>% select(review_id, score) %>%
group_by(review_id) %>% summarise(bing_score = sum(score))

afinn_aggregate <- afinn_sentiment %>% select(review_id, score) %>%
group_by(review_id) %>%  summarise(afinn_score = sum(score))
```

Next, we aggregate to the review level and add the sentiment information to our original imdb dataset. We include one column for each lexicon to make it easy to judge which scored the review sentiments best. Next, we use the merge function to sequentially combine the aggregated lexicon information with the original imdb review data. Figure 13.7 illustrates how the merge function is used. In our case, we tell R to combine observations in the original imdb data with the aggregated lexicon data based on the matching review_id. By specifying all.x = TRUE, merge is acting as a left-join operator, because we specify that we want to keep all of the observations in imdb and we want to add the aggregated lexicon score only if we find a match between the review_ids in the two dataframes. Then, we replace NAs that arise from non-matches with 0 s, indicating a neutral sentiment score.

```
imdb_sent <- merge(x = imdb, y = nrc_aggregate, all.x = TRUE,
by = "review_id")
imdb_sent <- merge(x = imdb_sent, y = bing_aggregate, all.x = TRUE,
by = "review_id")
imdb_sent <- merge(x = imdb_sent, y = afinn_aggregate, all.x = TRUE,
by = "review_id")
imdb_sent[is.na(imdb_sent)] <- 0
```

Finally, we use the ifelse function again to denote if the review is positive or negative, according to each of the three lexicons. These sentiment flags are added as

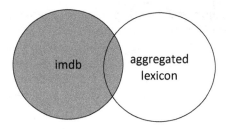

Fig. 13.7 Diagram of a left join between imdb and the aggregated lexicons

new columns to the imdb data and saved as a dataframe called imdb_sent. We use the same procedure to denote the actual sentiment of the review in words in a column called "actual sentiment."

```
imdb_sent$afinn_judgement <- ifelse(imdb_sent$afinn_score < 0, "negative",
                               ifelse(imdb_sent$afinn_score > 0, "positive",
                               "neutral"))
imdb_sent$bing_judgement <- ifelse(imdb_sent$bing_score < 0, "negative",
                               ifelse(imdb_sent$bing_score > 0, "positive",
                               "neutral"))
imdb_sent$nrc_judgement <- ifelse(imdb_sent$nrc_score < 0, "negative",
                               ifelse(imdb_sent$nrc_score > 0, "positive",
                               "neutral"))
Imdb_sent$actual_sentiment <- ifelse(imdb_sent$positive_flag == 1, "positive",
                               "negative")
```

13.6   Sentiment Analysis Results

Using the sentiment flags, we can compare the lexicon-based sentiments with the actual labeled sentiments of the reviews using the table function. First, we can compare the predicted and actual sentiments using the NRC lexicon.

```
table(imdb_sent$nrc_judgement, imdb_sent$actual_sentiment)

          negative positive
negative      235       42
neutral       160      166
positive      105      292
```

The NRC lexicon correctly identifies 292 positive reviews, or 58.4%, and 235 negative reviews, or 47%. Overall, based on the NRC lexicon, approximately 53% of the reviews have the correct sentiment.

```
table(imdb_sent$bing_judgement, imdb_sent$actual_sentiment)

          negative positive
negative     304      43
neutral      119     110
positive      77     347
```

The bing lexicon correctly identified 304 of the 500 (60.8%) negative reviews and 347 of the 500 (69.4%) positive reviews. It did substantially better for the negative reviews and showed an improvement over the NRC lexicon for the positive reviews. Overall, it classified 651 (69.4%) of the 1,000 reviews correctly. Finally, we consider the accuracy of the AFINN lexicon.

```
table(imdb_sent$afinn_judgement, imdb_sent$actual_sentiment)

          negative positive
negative     274      35
neutral      133     127
positive      93     338
```

AFINN correctly identified 274 of the 500 (54.8%) negative reviews and 338 of the 500 (67.6%) positive reviews. Overall, 612 of the 1,000 (61.2%) were labeled correctly.

Based on the results for the three lexicons, we would choose the bing lexicon to identify the sentiments of the movie reviews. If we were to acquire additional imdb reviews with unknown labels, we would expect the bing lexicon to produce the best results. We can dig deeper into the results by looking at the misclassified reviews. How many reviews did all three lexicons classify incorrectly?

```
incorrect_reviews <- imdb_sent[imdb_sent$afinn_judgement !=
               imdb_sent$actual_sentiment &
               imdb_sent$bing_judgement != imdb_sent$actual_sentiment &
               imdb_sent$nrc_judgement != imdb_sent$actual_sentiment,]
nrow(incorrect_reviews)
[1] 205
```

There are 205 reviews that all three lexicons predicted incorrectly. It can be insightful to read a sample of those misclassified reviews by running the following code. Five of these consistently misclassified reviews are shown in Table 13.1 with the actual sentiment and the predicted sentiment for the three lexicons.

```
sample <- incorrect_reviews[sample(nrow(incorrect_reviews), 5), c("comment",
               "actual_sentiment", "afinn_judgement", "bing_judgement",
               "nrc_judgement")]
```

Table 13.1 Misclassified sentiment reviews with actual and predicted sentiment

Comment	actual_ sentiment	afinn_ judgement	bing_ judgement	nrc_ judgement
Tom Wilkinson's character is a man who is not prepared for the ordeal that is about to begin, but he takes the matter in hand as the story progresses, and this great actor gives a performance that makes you feel the character's anguish and suffering	Positive	Neutral	Negative	Negative
Lots of holes in the script	Negative	Neutral	Neutral	Positive
None of them are engaging or exciting	Negative	Positive	Positive	Positive
Overall, I rate this movie a 10 out of a 1–10 scale	Positive	Neutral	Neutral	Neutral
There are massive levels, massive unlockable characters… it's just a massive game	Positive	Neutral	Neutral	Neutral

As described in Chap. 10, sentiment analysis struggles to correctly classify the sentiment of ambiguous text. Given the context, a human reader would know that the phrase "Lots of holes in the script" is negative. However, sentiment analysis is context-free and therefore incorrectly identifies this ambiguous review. For those reviews that are ambiguous to human readers, such as "There are massive levels, massive unlockable characters… it's just a massive game," sentiment analysis also has difficulty identifying the sentiment.

Additionally, our sentiment analysis fails to account for negation, as in the review "None of them are engaging or exciting." In this case, sentiment analysis identifies *engaging* and *exciting* as positive and deems the review positive, even though to a human reader the sentiment is clearly negative.

13.7 Custom Dictionary

The general dictionaries are a good start, but they were not created with analysis-specific context in mind and leave important terms out as a result. One way to improve our sentiment analysis results is to create a dictionary that is specific to our analysis. In creating the dictionary, we need to achieve a balance between improvement in predictive performance and generalizability to out-of-sample data. A dictionary that is too specific will have strong predictive accuracy for the data on which the model is built, the training data, but may perform poorly on our testing data. A dictionary that is too broad will miss information.

To create a custom dictionary, we split our data into a training and test set, as in Chap. 9, covering classification analysis. The test set provides an estimate of how our new model would perform on an out-of-sample dataset. Our goal is to create a dictionary that improves the predictive performance on the test set compared to the bing lexicon's performance.

To accomplish this goal, we return to the original imdb dataframe. We use the sample function on the imdb dataset to create the two samples, training and testing.

The sample function will provide us with the row numbers of 75% of our imdb data. In our analysis, the training set will be 75% of our data, and the test set will have 25%. We will split our data into the two samples based on the imdb data and then subset our tidy-formatted data to avoid preprocessing the data again. We use the head function to view the first six observations. The head function shows that Review 1 is in our training sample.

```
set.seed(101)
train <- sample(nrow(imdb), nrow(imdb)*.75)
tidy_train <- subset(tidy_imdb, review_id %in% train)
tidy_test <- subset(tidy_imdb, !(review_id %in% train))
head(tidy_train)
    positive_flag review_id    word
1               0         1       a
1.1             0         1    very
1.2             0         1    very
1.3             0         1    very
1.4             0         1    slow
1.5             0         1  moving
```

We now need a dictionary for our newly created sentiments. To create the custom dictionary, we begin with the best performing dictionary, bing, and add to it based on data exploration. We use left_join to add the bing lexicon sentiments to the tidy_train data.

```
tidy_train_imdb <- tidy_train %>% left_join(get_sentiments("bing"),
                   by = "word")
colnames(tidy_train_imdb)[which(colnames(tidy_train_imdb) == "sentiment")]
                   <-"bing_sentiment"
tidy_train_imdb$bing_score <- ifelse(is.na(tidy_train_imdb$bing_sentiment), 0,
                             ifelse(tidy_train_imdb$bing_sentiment
                             == "positive", 1, -1))
head(tidy_train_imdb)
  positive_flag review_id    word bing_sentiment bing_score
1             0         1       a          <NA>          0
2             0         1    very          <NA>          0
3             0         1    very          <NA>          0
4             0         1    very          <NA>          0
5             0         1    slow      negative         -1
6             0         1  moving          <NA>          0
```

To start, we look at the words in the imdb reviews that have zero polarity according to the bing lexicon. Let's check the words that appear most in the negative and positive reviews but do not have a polarity. First, we calculate term frequency and document frequency, as defined in Chap. 5, for those terms that are not classified by the bing lexicon. We create a dataframe and sort it in descending order by total term usage. Given the six most frequent terms, we may want to reconsider our decision not to remove stop words.

```
zero_polar <- tidy_train_imdb %>%
   filter(is.na(bing_sentiment) %>%
   select(positive_flag, review_id, word) %>%
   group_by(word) %>%
   summarise(total_usage = n(),
             review_frequency = n_distinct(review_id)) %>%
   arrange(desc(total_usage))
head(zero_polar)
# A tibble: 6 x 3
   word total_usage review_frequency
   <chr>        <int>             <int>
1    the          649               381
2      a          332               253
3    and          320               243
4     of          284               230
5     is          261               227
6   this          214               199
```

Let's remove the stop words that do not exist in the bing lexicon and repeat the steps to create zero_polar.

```
stop_words <- stop_words %>% anti_join(get_sentiments("bing"), by = "word")
tidy_train_imdb <- tidy_train_imdb %>% anti_join(stop_words, by = "word")
zero_polar <- tidy_train_imdb %>%
   filter(is.na(bing_sentiment)) %>% #select words that bing did not review
   select(positive_flag, review_id, word) %>%
   group_by(word) %>%
   summarise(total_usage = n(),
             review_frequency = n_distinct(review_id)) %>%
   arrange(desc(total_usage))
head(zero_polar)
# A tibble: 6 x 3
        word total_usage review_frequency
        <chr>        <int>             <int>
1      movie          131               123
2       film          128               124
3       time           38                38
4     acting           33                33
5 characters           26                26
6         10           25                17
```

This process improves the results. Next, we use the same procedure to calculate the frequency of the unclassified terms in the labeled reviews to explore term frequency by known sentiment label. We create two dataframes, zero_polar_positive and zero_polar_negative, which include the total frequency of the terms and the total number of reviews in which the terms appear in positive and negative-labeled reviews.

```
zero_polar_positive <- tidy_train_imdb %>%
    #select words that bing did not review and are part of a positive review
    filter(is.na(bing_sentiment) & positive_flag == 1) %>%
    select(review_id, word) %>%
    group_by(word) %>%
    summarise(positive_total_usage = n(),
            positive_review_frequency = n_distinct(review_id)) %>%
    arrange(desc(positive_total_usage))

zero_polar_negative <- tidy_train_imdb %>%
    #select words that bing did not review and are part of a negative review
    filter(is.na(bing_sentiment) & positive_flag == 0) %>%
    select(review_id, word) %>%
    group_by(word) %>%
    summarise(negative_total_usage = n(),
            negative_review_frequency = n_distinct(review_id)) %>%
    arrange(desc(negative_total_usage))
```

Again, we use the merge function to complete a left join adding the positive and negative term review information from the zero_polar_positive and zero_polar_negative dataframes to the zero_polar dataframe. Then, we add a column to the dataframe, percent_positive_review, which includes the calculated percentage of positive reviews in which the term appears. We replace any NAs arising from the left join with 0 and sort the zero_polar dataframe in descending order by review_frequency and percent_positive_review.

```
zero_polar <- merge(x = zero_polar, y = zero_polar_negative, all.x = TRUE,
                by = "word")
zero_polar <- merge(x = zero_polar, y = zero_polar_positive, all.x = TRUE,
                by = "word")
zero_polar[is.na(zero_polar)] <- 0
zero_polar$percent_positive_review <- round((zero_polar$positive_review_freque
ncy / zero_polar$review_frequency),4)
zero_polar <- zero_polar %>% arrange(desc(review_frequency),
            desc(percent_positive_review))
```

Next, we use the head function to view the 10 most frequent words. As we suspected, the most frequent terms appear to be stop words.

```
zero_polar %>% head(10) %>% select(word, review_frequency,
percent_positive_review)
          word review_frequency percent_positive_review
1         film              124                  0.6452
2        movie              123                  0.5203
3         time               38                  0.3684
4       acting               33                  0.4545
5   characters               26                  0.6538
6       movies               23                  0.5652
7     watching               19                  0.5263
8           10               17                  0.8235
9        films               17                  0.7059
10   character               17                  0.5882
```

Looking at the ten most frequently uncategorized words, most of them are words we would expect from reviews about movies: *film, movie, characters*, etc. There are several words that account for large percentages in positive or negative reviews—*10, films, time*. Reviews with *10* present are positive 82.4% of the time. This high rate intuitively makes sense—"This movie was a 10!" The words *time* and *films* are more ambiguous. In the 38 reviews where *time* is present, they are negative in 36.8% of the cases. Reviews with *films* are positive 70.6% of the time. Let's check the usage of *10, films*, and *time*. First, we will look at *10*. We subset the imdb data by the reviews that include *10*. The reviews and their sentiment labels are displayed in Table 13.2.

```
example_ids <- tidy_train_imdb %>% filter(word == "10") %>% select(review_id)
   %>% unique() %>% arrange(review_id)
imdb[imdb$review_id %in% example_ids$review_id, "comment"]
```

Most are clearly positive. There are two reviews that list "1/10" or "1 out of 10," but we will accept the errors on those reviews for the correctly classified positive reviews. Next, we look at the reviews that use the word *time* (Table 13.3).

```
example_ids <- tidy_train_imdb %>% filter(word == "time") %>%
             select(review_id) %>% unique() %>% arrange(review_id)
imdb[imdb$review_id %in% example_ids$review_id, c("comment", "positive_flag")]
```

The word *time* is less clear-cut. *Time* has many uses, and we see it is used positively and negatively. Reading the reviews to check their validity is tedious and does not give us an idea of the change in classification performance by adding terms to the bing lexicon. To make it easier to check, let's create a function that will output

Table 13.2 Reviews with *10*, including review number, text, and sentiment labels

Review	Review text	positive_flag
62	All in all I give this one a resounding 9 out of **10**	1
82	This is the first movie I've given a **10** to in years	1
87	I gave it a **10**	1
126	**10/10**	1
292	Rating: 1 out of **10**	0
302	**10** out of **10** for both the movie and trilogy	1
357	Now you know why I gave it a **10+**	1
435	1/**10**—and only because there is no setting for 0/**10**	0
439	Still, it was the SETS that got a big "**10**" on my "oy-vey" scale	1
449	My 8/**10** score is mostly for the plot	1
487	Overall I rate this movie a **10** out of a 1–**10** scale	1
523	Rating: 0/**10** (grade: Z) note: The show is so bad that even the mother of the cast pull her daughter out of the show	0
633	I'll give this film **10** out of **10**	1
653	Don't be afraid of subtitles........ it's worth a little aversion therapy **10/10**	1
688	**10** out of **10** stars	1
789	**10/10**	1
931	I rate this movie 9/**10**	1

Table 13.3 Reviews with *time*, including review number, text, and sentiment labels

Review	Review text	Positive_flag
34	Actually, the graphics were good at the **time**	1
75	I wouldn't say they're worth 2 h of your **time**, though	0
85	Plus, it was well-paced and suited its relatively short run **time**	1
99	Ursula Burton's portrayal of the nun is both touching and funny at the same **time** without making fun of nuns or the church	1
138	IMDB ratings only go as low as 1 for awful; it's **time** to get some negative numbers in there for cases such as these	0
143	It is an hour and half waste of **time**, following a bunch of very pretty high schoolers whine and cry about life	0
217	Ironically I mostly find his films a total waste of **time** to watch	0
235	Do not waste your **time**	0
293	An AMAZING finale to possibly the BEST trilogy of all **time**	1
332	For those that haven't seen it, don't waste your **time**	0
338	By the **time** the pyromaniac waylaid the assistant, I was bored and didn't care what happened next, and so I switched off	0
416	As a European, the movie is a nice throwback to my **time** as a student in the 1980s and the experiences I had living abroad and interacting with other nationalities, although the circumstances were slightly different	1
432	I am not a filmmaker nor am I a director, but I would hide my head in the sand if I'd spent whatever amount of money and **time** on this movie	0
433	In short, this was a monumental waste of **time** and energy, and I would not recommend anyone to EVER see this film	0
450	I won't say anymore—I don't like spoilers, so I don't want to be one, but I believe this film is worth your **time**	1
508	By the **time** the film ended, I not only disliked it, but I despised it	0

(continued)

Table 13.3 (continued)

Review	Review text	Positive_flag
589	Even when the women finally show up, there is no sign of improvement; the most expected things happen, and by the **time** the film is over, you might be far asleep	0
593	Being a 1990s child, I truly enjoyed this show, and I can proudly say that I enjoyed it big **time** and even more than the classical WB cartoons	1
611	Don't waste your **time** watching this rubbish non-researched film	0
665	Even allowing for poor production values for the **time** (1971) and the format (some kind of miniseries), this is baaaaaad	0
700	Very bad performance played by Angela Bennett, a computer expert who is at home all the **time**	0
706	One of the worst shows of all **time**	0
710	Every **time** he opened his mouth, you expect to hear, "you see kids…" Pulling the plug was a mercy killing for this horrible show	0
719	Top line: don't waste your **time** and money on this one; it's as bad as it comes	0
725	Seriously, it's not worth wasting your or your kid's **time** on	0
755	It is rare when a filmmaker takes the **time** to tell a worthy moral tale with care and love that doesn't fall into the trap of being overly syrupy or overindulgent	1
782	The foreigner is not worth 1 s of your **time**	0
823	It's a long **time** since I was so entertained by a movie	1
826	It's pretty surprising that this wonderful film was made in 1949, as Hollywood generally had its collective heads in the sand concerning black and white issues at that **time**	1
828	Plus, with the movie's rather modest budget and fast running **time**, it does an amazing job	1
838	Don't waste your **time**	0
857	It's dumb and pointless and a complete waste of **time**	0
897	I felt asleep the first **time** I watched it, so I can recommend it for insomniacs	0
901	Otherwise, don't even waste your **time** on this	0
928	It was a long **time** that I didn't see a so charismatic actor on screen	1
930	The movie is not completely perfect, but "Titta Di Girolamo" will stay with you for a long **time** after the vision of the movie	1
970	At a **time** when it seems that film animation has been dominated by Disney/Pixar's CGI masterpieces, it is both refreshing and comforting to know that Miyazaki is still relying on traditional hand-drawn animation to tell his charming and enchanting stories	1
975	If you haven't choked in your own vomit by the end (by all the cheap drama and worthless dialogue), you've must have bored yourself to death with this waste of **time**	0

the change in correctly classified reviews by classifying new terms that are not classified by the bing lexicon. Custom functions in R can be created by using "function." The parameters that need to be specified are listed in parentheses. For our custom function, called check_word, we need to provide a word and a sentiment as inputs to the function.

```
check_word <- function(input_word, sentiment_input){
  if(!(sentiment_input %in% c("negative", "positive")))
  stop("sentiment must be 'positive' or 'negative'")
  bing_edit_check <- rbind(bing_edit, c(input_word, sentiment_input))
  check_tidy_df <- tidy_train_imdb %>% left_join(bing_edit_check,
                       by = "word")
  check_tidy_df$new_score <- ifelse(is.na(check_tidy_df$sentiment), 0,
                           ifelse(check_tidy_df$sentiment == "positive",
                           1, -1))
  check_agg_score <- check_tidy_df %>% select(review_id, positive_flag,
                       bing_score, new_score) %>%
                       group_by(review_id, positive_flag) %>%
                       summarise(bing_score = sum(bing_score),
                       new_score = sum(new_score)) %>%
                       mutate(bing_judge = ifelse(bing_score > 0, "positive",
                              ifelse(bing_score < 0, "negative", "neutral"),
                       edit_judge = ifelse(new_score > 0, "positive",
                              ifelse(new_score < 0, "negative", "neutral")))
  new_score <- c(positive = nrow(check_agg_score[check_agg_score$edit_judge
             == "positive" & check_agg_score$positive_flag == 1, ]),
                 negative = nrow(check_agg_score[check_agg_score$edit_judge
             == "negative" & check_agg_score$positive_flag == 0, ]))
  new_score <- c(new_score, total_correct = sum(new_score))
  bing_score <- c(positive = nrow(check_agg_score[check_agg_score$bing_judge
             == "positive" & check_agg_score$positive_flag == 1, ]),
                 negative = nrow(check_agg_score[check_agg_score$bing_judge
             == "negative" & check_agg_score$positive_flag == 0, ]))
  bing_score <- c(bing_score, total_correct = sum(bing_score))
  rbind(new_scores = new_score, bing_scores = bing_score)
}
```

By using the check_word function, we can compare the sentiment classification performance without *10* classified by sentiment with the reviews including *10* as a positive sentiment term.

```
check_word("10","positive")
            positive negative total_correct
new_scores       283      221           504
bing_scores      271      222           493
```

Based on the output of the function, by adding *10* as a positive sentiment term, we increase the total correctly classified reviews by 11. Next, we try our function with the term *time* as a negative word.

```
check_word("time", "negative")
           positive negative total_correct
new_scores      264      224           488
bing_scores     271      222           493
```

Based on the output of our function, this change reduces our accuracy. This result makes sense, based on the reviews with the word. It is used in many ways, and the negative reviews typically contain the word *waste* nearby. For this reason, we do not add *time* to our custom dictionary. Finally, we consider adding the term *films* as a positive sentiment term.

```
check_word("films", "positive")
           positive negative total_correct
new_scores      273      219           492
bing_scores     271      222           493
```

Based on the results of the function, the word *films* results in a relatively unchanged score, so we will not add it to our custom dictionary. Now, we need to look for additional words to test. Having reviewed the most frequent 10 words, let's look at words 11 through 30. We must be careful, because words in this group have lower frequencies. Therefore, we are more likely to see a high or low positive percent just by chance. For this reason, large datasets are preferred in sentiment analysis.

```
zero_polar[11:30, c("word", "review_frequency", "percent_positive_review")]
              word review_frequency percent_positive_review
11           story               17                  0.4706
12            real               16                  0.5625
13            cast               14                  0.7857
14          actors               14                  0.6429
15           watch               14                  0.5000
16          script               14                  0.2143
17          scenes               13                  0.4615
18         writing               11                  0.3636
19             art               10                  0.9000
20          people               10                  0.8000
21          screen               10                  0.7000
22            life               10                  0.6000
23          totally               10                  0.3000
24          played                9                  1.0000
25     performance                9                  0.8889
26           music                9                  0.7778
27           short                9                  0.7778
28             job                8                  1.0000
29            play                8                  1.0000
30           actor                8                  0.8750
```

None of these words seem to be indicative of a positive or negative sentiment, but we can try testing them using our check_words function. First, we isolate the reviews and their sentiments. Then, we use lapply to apply the function to the terms.

```
words_to_test <- zero_polar[11:30, c("word", "percent_positive_review")]
words_to_test$polarity <- ifelse(words_to_test$percent_positive_review < .5,
                       "negative", "positive")
words_test <- lapply(1:nrow(words_to_test), function(x)
          check_word(words_to_test[x, "word"],
          words_to_test[x,"polarity"]))
names(words_test) <- words_to_test$word
words_test
$story
          positive negative total_correct
new_scores      267      222           489
bing_scores     271      222           493

$real
          positive negative total_correct
new_scores      274      218           492
bing_scores     271      222           493

$cast
          positive negative total_correct
new_scores      271      221           492
bing_scores     271      222           493

$actors
          positive negative total_correct
new_scores      273      221           494
bing_scores     271      222           493

$watch
          positive negative total_correct
new_scores      272      221           493
bing_scores     271      222           493

$script
          positive negative total_correct
new_scores      269      225           494
bing_scores     271      222           493

$scenes
          positive negative total_correct
new_scores      269      222           491
bing_scores     271      222           493
```

```
$writing
           positive negative total_correct
new_scores      269      222            491
bing_scores     271      222            493

$art
           positive negative total_correct
new_scores      272      222            494
bing_scores     271      222            493

$people
           positive negative total_correct
new_scores      274      222            496
bing_scores     271      222            493

$screen
           positive negative total_correct
new_scores      274      221            495
bing_scores     271      222            493

$life
           positive negative total_correct
new_scores      273      220            493
bing_scores     271      222            493

$totally
           positive negative total_correct
new_scores      271      223            494
bing_scores     271      222            493

$played
           positive negative total_correct
new_scores      272      222            494
bing_scores     271      222            493

$performance
           positive negative total_correct
new_scores      272      221            493
bing_scores     271      222            493

$music
           positive negative total_correct
new_scores      272      222            494
bing_scores     271      222            493
```

```
$short
            positive negative total_correct
new_scores       271      222            493
bing_scores      271      222            493


$job
            positive negative total_correct
new_scores       272      222            494
bing_scores      271      222            493


$play
            positive negative total_correct
new_scores       274      222            496
bing_scores      271      222            493


$actor
            positive negative total_correct
new_scores       271      221            492
bing_scores      271      222            493
```

Looking through the results, we notice that including these terms in our custom dictionary generally impacts our correctly classified score by only 1 or 2. The maximum increase occurs when we include the terms *people* and *play*, but they increase the correctly classified reviews only by 3. We could continue looking for words to improve our dictionary in this manner. In the interests of brevity, we will continue with our findings about classifying *10* as a positive sentiment term.

13.8   Out-of-Sample Comparison

We will use the test set to test the bing lexicon against our edited bing lexicon. We need to repeat the same steps as before on the test set with both dictionaries.

```
tidy_test_imdb <- tidy_test %>%
  left_join(bing_edit, by = "word") %>%
  left_join(get_sentiments("bing"), by = "word") %>%
  mutate(edit_score = ifelse(is.na(sentiment_edit), 0,
                       ifelse(sentiment_edit == "positive", 1, -1)),
         standard_score = ifelse(is.na(sentiment), 0,
                       ifelse(sentiment == "positive", 1, -1))) %>%
  group_by(positive_flag, review_id) %>%
  summarise(edit_score = sum(edit_score),
            standard_score = sum(standard_score)) %>%
  mutate(edit_value = ifelse(edit_score > 0, "positive",
                       ifelse(edit_score < 0, "negative", "netural")),
         standard_value = ifelse(standard_score > 0, "positive",
                       ifelse(standard_score < 0, "negative", "neutral")))
```

First, we look at the accuracy of the sentiment analysis based on the original bing lexicon.

```
table(tidy_test_imdb$positive_flag, tidy_test_imdb$standard_value)

    negative neutral positive
0         82      34       21
1          9      28       76
```

On the test set, the regular bing lexicon correctly predicted 82 of the 137 negative reviews (59.85%) and 76 of the 113 positive reviews (67.25%), for an overall result of 158 of the 250 correct (63.2%). Next, we look at the accuracy of the sentiment analysis based on the custom lexicon.

```
table(tidy_test_imdb$positive_flag, tidy_test_imdb$edit_value)

    negative netural positive
0         82      31       24
1          9      27       77
```

The edited lexicon correctly scored 82 of the 137 negative reviews (59.85%), just as the regular lexicon did. The correctly scored positive reviews increased by 1–77 of the 113 (68.1%). Overall, 159 of the 250 were scored correctly (63.6%). These results do not indicate any significant change. Let's check the use of *10* in the reviews.

```
reviews_10 <- tidy_test  %>% filter(word == "10") %>% select(review_id)
imdb[imdb$review_id %in% reviews_10$review_id, c("comment", "positive_flag")]
comment
331 The hockey scenes are terrible, defensemen playing like they're 5 years
old, goalies diving at shots that are 10 feet wide of the net, etc.
354 I would give this television series a 10 plus if i could.
459 This gets a 1 out of 10, simply because there's nothing lower.
617 My rating: just 3 out of 10.
    positive_flag
331             0
354             1
459             0
617             0
```

As we can see, there are only four reviews that contain *10* in the test set, which is one of the disadvantages of using a small dataset. Three of them had a negative original review but were classified as neutral by the bing lexicon. Based on the out-of-sample results, our edited dictionary appears to perform similarly to the original bing lexicon. This outcome is not surprising, because we only made one change. For improved accuracy, we could consider switching to n-grams and weeding out the use of *10* in cases such as "out of 10."

Key Takeaways
- R open-source software, more specifically RStudio IDE, can be used to perform lexicon-based sentiment analysis using online consumer reviews.
- Built-in lexicons can easily be used and manipulated, and custom stop lists can be created in R.

References

Hu, M., & Liu, B. (2004, August). Mining and summarizing customer reviews. In *Proceedings of the Tenth ACM SIGKDD International Conference on Knowledge Discovery and Data Mining* (pp. 168–177). ACM.

Mohammad, S. M., & Turney, P. D. (2013). Crowdsourcing a word–emotion association lexicon. *Computational Intelligence, 29*(3), 436–465.

Nielsen, F. Å. (2011). A new ANEW: Evaluation of a word list for sentiment analysis in microblogs. arXiv preprint arXiv:1103.2903.

R Development Core Team. (2008). *R: A language and environment for statistical computing.* Vienna: R Foundation for Statistical Computing. ISBN 3-900051-07-0, URL http://www.R-project.org

Further Reading

For more about R software, see R Development Core Team (2008) and visit https://www.r-project.
org/

Chapter 14
Latent Semantic Analysis (LSA) in Python

Abstract This chapter presents the application of latent semantic analysis (LSA) in Python as a complement to Chap. 6, which covers semantic space modeling and LSA. In this chapter, we will present how to implement text analysis with LSA through annotated code in Python. In this example, we will run LSA over a dataset that includes 401 instances of both online and offline review sources from the Areias do Seixo Eco-Resort (Data available at https://archive.ics.uci.edu/ml/data-sets/Eco-hotel).

Keywords Python · Latent semantic analysis · Text analytics · Text mining

14.1 Introduction to Python and IDLE

Similar to R, which is presented in Chap. 13, Python offers the advantages of open-source software, such as being free and offering countless—also free—online resources to develop and improve the code and the analysis itself. Some tools allow R users to write code for Python and vice versa. For example, Python users can develop R code by using rpy. Nevertheless, generally speaking, users may need to choose one language or the other.

These two languages differ in several aspects. R is best suited for statistics and data exploration. Python is a more generic programming language than R, and it is more approachable and intuitive than other software development languages. For those users interested in learning a general language that can do more than mathematics and statistics, Python is preferable. It is also especially well suited for data manipulation and repeated, routine tasks. Arguably, Python is easier to learn than R, which is considered to have a steeper learning curve. A final consideration is the underlying logic of these two languages. Python programming follows the logic of how computer programmers think. R programming follows a logic closer to how statisticians and mathematicians think. Therefore, R programmers may find Python's programming somewhat difficult and counterintuitive and vice versa.

Another unique characteristic of Python is that it has two versions concurrently updated—Python 2.X and Python 3.X. We recommend that readers use the Python 3.X version. While some of the resources currently available are based only on Python 2.X and some libraries do not support Python 3.X, Python has scheduled the discontinuation of the 2.X versions several times. The last date provided for this discontinuation is 2020. Therefore, this chapter will use the Python 3.X version.

14.2   Preliminary Steps

If you do not have Python installed on your computer, you can download the platform-specific version at: https://www.python.org/downloads/. Since Python is platform-dependent, we refer you to the beginners guide available at: https://wiki.python.org/moin/BeginnersGuide/Download.[1] While the implementation may vary depending on the platform, the code will generally be consistent. This chapter uses Python 3.6 on a Windows machine.

We begin by installing the necessary modules. In our case, we will use several modules, including *csv*, *nltk*, *pandas*, *numpy*, *sklearn*, *re*, and *matplotlib*. The *csv* module allows Python to read and write tabular data in comma separated values (.csv) format. The *nltk* module (nltk stands for Natural Language Toolkit) allows us to use several text processing and preprocessing functions, such as parsing, tokenization, stemming, classification, and tagging. The *pandas* module provides structure to the data, which simplifies its analysis and manipulation. The Python module *NumPy* provides several generic functions for scientific computing. In our analysis, *NumPy* is used as a multidimensional container of our review data. *Sklearn* incorporates a set of different modules into our analyses for data mining and machine learning. The module *re* provides matching functions for 8-bit and Unicode strings. *Matplotlib* allows us to generate high-quality plots, charts, and histograms.

In our Windows-based implementation, we use the command line to install the modules, with our path as the location of the Python scripts folders. We use pip for the installation.

```
C:\>cd C:\Python\Python36\Scripts
C:\Python\Python36\Scripts>pip install csv
C:\Python\Python36\Scripts>pip install nltk
C:\Python\Python36\Scripts>pip install pandas
C:\Python\Python36\Scripts>pip install numpy
C:\Python\Python36\Scripts>pip install sklearn
C:\Python\Python36\Scripts>pip install re
C:\Python\Python36\Scripts>pip install matplotlib
```

---

[1] Additional resources are available for nonprogrammers (https://wiki.python.org/moin/BeginnersGuide/NonProgrammers) or programmers (https://wiki.python.org/moin/BeginnersGuide/Programmers).

In this chapter, we will run our code in the Python integrated development and learning environment (IDLE), which should be automatically downloaded when you install Python. This interactive environment serves a similar purpose as RStudio with R, presented in Chap. 13. We use Python IDLE, because according to its documentation, it works consistently across computer platforms. Alternatively, .py files can be run directly in the command window, after using cd to specify the directory containing the .py file. For example, to run a file called file.py, which is

```
C:\Python\Python36\Scripts>cd C:\
C:\>python file.py
```

Once the Python IDLE is open, the default behavior opens the Shell. In the Shell, we can run individual lines of code. The initial Shell file is depicted in Fig. 14.1.

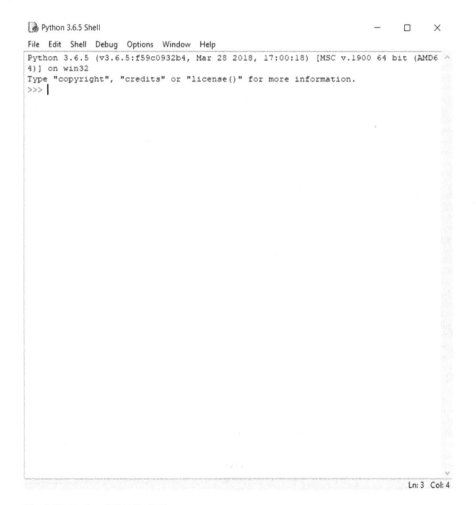

Fig. 14.1 Python IDLE Shell file

Fig. 14.2 Python IDLE script and .py file

Upon opening the IDLE Shell, we can either choose to open a .py file or use the File menu to create a New File. Choosing New File will open another window, which will be blank. In order to run the new file, we can use the File menu to choose Run and then Run Module. An example of a .py file and the Shell are depicted in Fig. 14.2, with the shell in the background and the file in the .py foreground. In choosing Run Module, the full .py file will run, and any necessary output will display in the Shell.

14.3 Getting Started

First, we define the problem or the reason for performing the analysis. In this example, we aim to uncover the latent content in our sample of 401 hotel reviews that were collected between January and August of 2015. A unique characteristic of this dataset is that it incorporates two different sources of data, offline (e.g., guest books) and online reviews (e.g., Tripadvisor.com), in one single file. We want to extract insights from reviews written by former guests of the Areias do Seixo Eco-Resort.

To begin in Python, we import the necessary modules or packages. Modules consist of the top-level package name and associated submodules. To import the top-level module, we use "import."

```
>>> import csv
>>> import nltk
>>> import sklearn
>>> import re
```

We can import modules and assign nicknames to them to make them easier to refer to in our code by using "import…as." For instance, instead of typing "pandas" to refer to the pandas module, we can assign it the name pd.

```
>>> import pandas as pd
>>> import numpy as np
>>> import matplotlib.pyplot as plt
```

We import submodules in a slightly different way. To download subroutines, we use "from…import." As shown in the first code line, we can also use "from…import *" to import all submodules from the top-level module pylab.

```
>>> from pylab import *
>>> from nltk.corpus import stopwords
>>> from pylab import *
>>> from sklearn.decomposition import TruncatedSVD
>>> from sklearn.feature_extraction.text import TfidfTransformer
>>> from sklearn.feature_extraction.text import CountVectorizer
>>> from sklearn.preprocessing import Normalizer
>>> from sklearn import metrics
>>> from sklearn.cluster import KMeans, MiniBatchKMeans
```

Now that we have loaded all of the modules that we will use, we need to download some additional nltk data and files by using nltk.download(). As Fig. 14.3 illustrates, an additional window will open when this line is run. Download all additional files by choosing "all."

```
>>> nltk.download()
```

14.4   Data and Data Import

Now that we have set up Python for our analysis, we can move on to our data. We want to import the data that we will be using in our analysis (note that the path to the location of the file must be specified). The data are contained in a single column of a .csv file. We use open() to open the "Data.csv" file, which we designate for

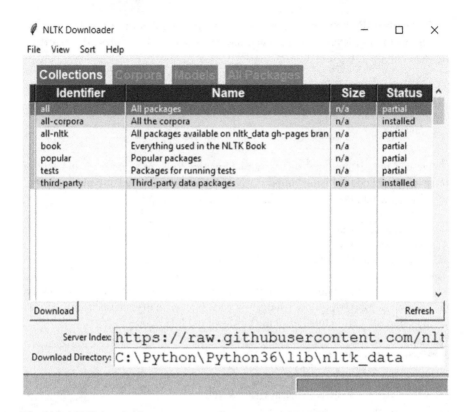

Fig. 14.3 NLTK downloader

read-only use by including the "r" argument. Just as we named our modules, we use "as" to assign the data the name f. We then use read.csv from the pandas module by using pd.read_csv () on our data, f, and setting it as "your_list."

```
>>> with open(' C:\\Users\\Data.csv','r') as f:
        your_list = pd.read_csv(f)
```

As described, the data contain reviews about a hotel. To begin, we may want to know more about the data that we imported. First, we can confirm the number of reviews using "len." As expected, we have 401 reviews. Then, we can find out more about the type of Python object we are working with using "type." As shown, your_ list is a pandas dataframe.

```
>>> len(your_list)
401
>>> type(your_list)
<class 'pandas.core.frame.DataFrame'>
```

We can preview the first and last five reviews by indexing the your_list dataframe. Unlike in R, in Python, indexing begins at 0. There are some convenient ways, however, to refer to the first and last observations in a list or dataframe. For instance, we can view the first five reviews in your_list using the code that follows. As displayed, the first five reviews are numbered from 0 to 4.

```
>>> your_list[:5]
                                    Review
0  Everything from the weather staff food propert...
1  The hotel it is fantastic built by the sea, li...
2  One dream! Cozy and comfortable Hotel! The bes...
3  Hotel concept is hard to grasp. They communica...
4  This is a wonderful hotel for a romantic escap...
```

We can view the last five reviews in the your_list dataframe using the code that follows. As displayed, the last five reviews are numbered from 396 to 400.

```
>>> your_list[-5:]
                                    Review
396  An extraordinary place! Amazing architecture h...
397  What a wonderful place to relax and enjoy the ...
398  Thank you for the best ever. And the best dinn...
399  As you know we have just returned home after 5...
400  SPA is excellent. I worked only with one lady ...
```

14.5 Analysis

Now that we are acquainted with the data, we begin our analysis by creating a function called fix_Text to preprocess the data. Prior to that step, we define stop_words, which is a set of English stop words. Then, we define the fix_Text function, which takes text as input. First, the function uses the *re* module to substitute a space for any special characters or numbers and saves the result as a string variable, letters_only. Then, we replace uppercase letters with lowercase letters and rename the result "words." There are some built-in functions for string manipulation, such as split, which splits the string based on a space. In other words, it separates each word. Next, for each of the words in our string called words, we compare the word to the stop_words and remove any matches that we find. We save the text that remains as meaningful. The result of our function is brought back together as processed strings.

```
>>> stop_words = set(stopwords.words("english"))
>>> def fix_Text(text):
        letters_only = re.sub("[^a-zA-Z]"," ", str(text))
        words=letters_only.lower().split()
        meaningful=[w for w in words if not w in stop_words]
        return(" ".join(meaningful))
```

We can look at the effect of our function on the first review and compare it to the original review. As we can see, the capital letter "E" was changed to lower case, and the following stop words were removed from the review: *from*, *the*, *and*, *were*.

```
>>> your_list["Review"][0]
'Everything from the weather staff food property fire pits decor spa rooms and beach were top notch'
>>> fix_Text(your_list["Review"][0])
'everything weather staff food property fire pits decor spa rooms beach top notch'
```

Next, we can iterate over all 401 reviews in our list of reviews to apply the function. First, we want to set the number of reviews as a named variable, which we call num_resp. Then, we create an empty object to store the result of the iteration, called clean_text. Finally, we iterate over our reviews to apply the function to each review in the list.

```
>>> clean_text = []
>>> for i in range(0,num_resp):
        clean_text.append(fix_Text(your_list["Review"][i]))
```

Now, we want to tokenize our reviews and create our document-term matrix. We use CountVectorizer, which creates the DTM based on raw frequency. We name our DTM object dtm and remove terms that do not appear more than once in the corpus.

```
>>> vectorizer = CountVectorizer(min_df=1)
>>> dtm = vectorizer.fit_transform(clean_text)
```

We then determine the dimensions of our DTM. Since we already know the number of documents, we use .shape to find out the number of terms in the DTM.

```
>>> dtm.shape
(401, 2305)
```

As shown, the DTM contains 401 reviews and 2,306 terms. Next, we can take a closer look at our DTM, which is labeled using the review number.

```
>>> pd.DataFrame(dtm.toarray(),index=range(0,401),
columns=vectorizer.get_feature_names()).head(10)

   able absolute absolutely absorb accept acceptable accommodating \
0   0     0         0         0      0      0           0
1   0     0         0         0      0      0           0
2   0     0         0         0      0      0           0
3   0     0         0         0      0      0           0
4   0     0         0         0      0      0           0
5   0     0         0         0      0      0           0
6   0     0         0         0      0      0           0
7   0     0         0         0      0      0           0
8   0     0         0         0      0      0           0
9   0     0         0         0      0      0           0

   accommodation account achieved ... years   yes yesterday yet \
0       0           0        0      ...   0     0     0        0
1       0           0        0      ...   0     0     0        0
2       0           0        0      ...   0     0     0        0
3       0           0        0      ...   0     0     0        0
4       0           0        0      ...   0     0     0        0
5       0           0        0      ...   0     0     0        0
6       0           0        0      ...   0     0     0        0
7       0           0        0      ...   0     0     0        0
8       0           0        0      ...   0     0     0        0
9       0           0        0      ...   0     0     0        0

   yiulyia  yoga young yuliya zen zero
0    0       0     0     0     0   0
1    0       0     0     0     0   0
2    0       0     0     0     0   0
3    0       0     0     0     0   0
4    0       0     0     0     0   0
5    0       0     0     0     0   0
6    0       0     0     0     0   0
7    0       0     0     0     0   0
8    0       0     0     0     0   0
9    0       0     0     0     0   0

[10 rows x 2306 columns]
```

As we can see, we are viewing the DTM for the first 10 reviews. We can view the first 10 terms using the following code.

```
>>> terms=vectorizer.get_feature_names()
>>> terms[0:10]
['able', 'absolute', 'absolutely', 'absorb', 'accept', 'acceptable', 'accommodating', 'accommodation',
'account', 'achieved']
```

Next, we can create a new DTM based on dtm, which uses *tfidf* weighting. We use TfidfTransformer to apply *tfidf* weighting to our DTM and save this weighted DTM as tfidf_dtm.

```
>>> tfidf_transformer = TfidfTransformer()
>>> tfidf_dtm = tfidf_transformer.fit_transform(dtm)
```

Now that we have created our weighted and unweighted DTMs, we use SVD for our LSA. In this case, we use TruncatedSVD. As arguments, we specify the number of latent factors and an algorithm. Since we have a large, sparse matrix, we will use the "randomized" algorithm. For our preliminary analysis, we use 25 as the number of factors. Later, we will create a scree plot to help us decide how many factors to ultimately use. To begin, we apply the LSA to our DTM and name the object created dtm_lsa. Finally, we normalize the resulting space.

```
>>> lsa = TruncatedSVD(25, algorithm='randomized')
>>> dtm_lsa = lsa.fit_transform(dtm)
>>> dtm_lsa = Normalizer(copy=False).fit_transform(dtm_lsa)
```

Next, we can view the LSA space for the first 10 latent factors, labeled 0 through 9. This view, which appears below, shows us the factor values for the first and last 10 terms in the LSA space.

```
>>> pd.DataFrame(lsa.components_, columns=vectorizer.get_feature_names()).head(10)
```

	able	absolute	absolutely	absorb	accept	acceptable \
0	0.013749	0.002820	0.039118	0.002550	0.001806	0.002556
1	0.007153	-0.002486	-0.012687	-0.003050	-0.003674	0.006149
2	0.001611	0.000862	0.014876	-0.001261	0.000052	0.001774
3	0.004097	-0.004221	0.015376	-0.003943	-0.001595	0.006665
4	-0.007110	-0.004201	-0.031583	0.002855	0.001609	-0.004062
5	-0.016659	-0.002920	-0.055943	-0.004642	0.000598	0.006568
6	-0.002070	0.003457	-0.055558	-0.002502	-0.001278	-0.017966
7	0.019146	0.004027	-0.020199	0.002751	0.006456	0.007479
8	-0.013958	0.003112	0.005797	-0.000551	0.000363	0.001963
9	-0.014358	0.000274	0.002453	0.011895	-0.000980	0.014423

	accommodating	accommodation	account	achieved	...	years \
0	0.004882	0.005554	0.011717	0.004470	...	0.023585
1	-0.011925	0.013095	0.007296	0.006829	...	0.073164
2	-0.002719	0.044802	0.005726	0.002927	...	0.283041
3	-0.007650	-0.008353	0.014277	0.013430	...	-0.105108
4	0.009967	0.021278	-0.019020	-0.000911	...	0.124869
5	0.007969	0.017900	-0.011214	-0.015462	...	0.025219
6	0.002784	-0.001202	-0.026749	-0.018708	...	-0.006957
7	0.018577	-0.000166	-0.007076	0.015968	...	-0.031582
8	-0.003164	-0.008399	-0.010897	0.000824	...	-0.029122
9	0.014313	-0.024550	-0.008660	0.000347	...	0.006237

	yes	yesterday	yet	yiulyia	yoga	young	yuliya \
0	0.007102	0.021952	0.019327	0.001578	0.017614	0.007249	0.004225
1	0.004620	0.074413	0.012693	0.002121	-0.003008	0.023884	-0.003464
2	0.002399	-0.050652	0.011927	0.000249	0.005494	0.089920	-0.005100
3	0.004801	-0.030493	0.021962	-0.000515	0.012503	-0.033139	-0.008973
4	-0.006877	0.040574	-0.045927	-0.002945	-0.018626	0.037674	0.008346
5	-0.002515	0.016911	0.054559	0.002198	-0.018786	0.009470	-0.002025
6	-0.020244	-0.031310	0.051893	0.002451	0.091475	-0.001698	-0.002950
7	0.007692	-0.017203	-0.009812	-0.000773	-0.007374	-0.014990	0.005530
8	0.000804	0.011918	0.024293	0.001056	-0.017515	-0.013066	-0.004551
9	0.018806	-0.017370	0.052262	-0.005253	0.012650	-0.002037	-0.005203

	zen	zero
0	0.001827	0.002979
1	-0.000200	0.004868
2	0.006600	0.001954
3	-0.003199	0.002298
4	0.005106	-0.000144
5	0.002329	-0.009377
6	-0.003140	0.000685
7	0.003255	0.007237
8	-0.008760	-0.001311
9	0.002598	0.014487

```
[10 rows x 2306 columns]
```

Next, we can view the first 10 reviews and the 25 factors.

```
>>> pd.DataFrame(dtm_lsa, index=range(0,401)).head(10)
          0         1         2         3         4         5         6 \
0  0.510987 -0.014657 -0.032065  0.111138 -0.110953 -0.043068 -0.109962
1  0.616895 -0.202628 -0.002306 -0.056770 -0.179124 -0.157713 -0.152674
2  0.583230  0.237838  0.160601  0.015477 -0.057968 -0.170911  0.120507
3  0.548577  0.161484  0.140848  0.178186 -0.232191 -0.115990 -0.137619
4  0.627850 -0.028411  0.168889  0.083308 -0.233225 -0.181913 -0.022517
5  0.386635 -0.233259  0.090502 -0.024438 -0.085955 -0.052243 -0.030432
6  0.565024 -0.049642  0.008636 -0.146977 -0.018576 -0.236534  0.172666
7  0.457475 -0.471364  0.011931 -0.228610  0.027218 -0.013820 -0.062061
8  0.605329 -0.062158  0.035770 -0.286809 -0.089867 -0.118332 -0.097990
9  0.352444 -0.378203 -0.028062 -0.160444  0.067359 -0.078973 -0.208093

          7         8         9    ...          15        16        17 \
0 -0.156421 -0.350032 -0.130046    ...     0.028201 -0.222759  0.254286
1 -0.035866  0.153570  0.125864    ...    -0.035328 -0.256524 -0.000390
2 -0.028251 -0.109417  0.082392    ...     0.317669 -0.028075  0.137788
3  0.253082  0.063792  0.003237    ...    -0.180549  0.012050  0.058678
4 -0.003614 -0.366153  0.044473    ...     0.138258 -0.115951  0.250343
5 -0.523088  0.075690 -0.275989    ...     0.395706  0.034237 -0.257962
6 -0.365759  0.145832 -0.042977    ...     0.106022  0.136114 -0.055560
7 -0.487943  0.081316 -0.174115    ...     0.156276 -0.062958 -0.274597
8 -0.256832  0.216131  0.099532    ...     0.021504 -0.031020  0.050775
9 -0.455523  0.357415 -0.083569    ...     0.140965  0.022662 -0.357771

         18        19        20        21        22        23        24
0 -0.350605  0.110179 -0.261564 -0.223285  0.032448  0.202093  0.202115
1 -0.119009 -0.130681  0.390618  0.094993  0.035315 -0.002806  0.134004
2  0.418444 -0.006926  0.259844 -0.020242 -0.044755 -0.114808  0.205581
3  0.075591  0.061278  0.069554  0.105619 -0.143809 -0.144725 -0.522724
4  0.089980  0.081567 -0.043008  0.075917  0.087638 -0.214700  0.111421
5 -0.156288  0.002638  0.119189 -0.034989 -0.063124  0.066788 -0.081429
6  0.087619  0.094267  0.181079 -0.188803 -0.010383 -0.130504  0.479287
7  0.038932 -0.023761 -0.174586  0.093763  0.039419 -0.141020 -0.054554
8  0.149981  0.058947 -0.063786 -0.198406 -0.079982  0.009973  0.433295
9  0.008779  0.031022  0.215526  0.051593  0.040296 -0.097362 -0.195113

[10 rows x 25 columns]
```

Next, we can calculate document similarity and view the first and last 10 documents.

```
>>> doc_sim=np.asarray(np.asmatrix(dtm_lsa)*np.asmatrix(dtm_lsa).T)
>>> pd.DataFrame(doc_sim, index=range(0,401), columns=range(0,401)).head(10)
          0         1         2         3         4         5         6  \
0  1.000000  0.321554  0.164317  0.097386  0.488586  0.217846   0.301747
1  0.321554  1.000000  0.480013  0.367089  0.448377  0.246754   0.488934
2  0.164317  0.480013  1.000000  0.285392  0.651149  0.143899   0.574373
3  0.097386  0.367089  0.285392  1.000000  0.382873  0.072660  -0.012807
4  0.488586  0.448377  0.651149  0.382873  1.000000  0.247993   0.333991
5  0.217846  0.246754  0.143899  0.072660  0.247993  1.000000   0.468969
6  0.301747  0.488934  0.574373 -0.012807  0.333991  0.468969   1.000000
7  0.185623  0.325449  0.092371  0.058211  0.309323  0.785270   0.457617
8  0.435776  0.504434  0.438371  0.116739  0.309950  0.353285   0.830504
9 -0.029467  0.521930  0.115535  0.152697  0.015523  0.663121   0.433980

          7         8         9    ...        391       392       393  \
0  0.185623  0.435776 -0.029467    ...   0.225699  0.156402  0.278194
1  0.325449  0.504434  0.521930    ...   0.439478  0.476610  0.289342
2  0.092371  0.438371  0.115535    ...   0.248294 -0.037384  0.133103
3  0.058211  0.116739  0.152697    ...  -0.003957  0.340499  0.001273
4  0.309323  0.309950  0.015523    ...   0.430351  0.226895  0.404533
5  0.785270  0.353285  0.663121    ...   0.163344  0.305761  0.179608
6  0.457617  0.830504  0.433980    ...   0.346683  0.042575  0.227739
7  1.000000  0.470497  0.767869    ...   0.627326  0.447842  0.543420
8  0.470497  1.000000  0.384236    ...   0.396954  0.124044  0.269111
9  0.767869  0.384236  1.000000    ...   0.452972  0.410034  0.337642

        394       395       396       397       398       399       400
0  0.157753  0.273576  0.604338  0.164714  0.415022  0.021273  0.164803
1  0.000523  0.006337  0.395841  0.373592  0.438532  0.232593  0.605098
2  0.012404  0.126345  0.105652  0.194226  0.217237  0.139543  0.345082
3  0.038839  0.052821  0.042930  0.066478 -0.110113  0.266072  0.248046
4  0.107073  0.105265  0.345977  0.307985  0.158374  0.271263  0.179939
5  0.108594  0.204301  0.397551  0.170228  0.359187  0.199241  0.293886
6  0.025676  0.125338  0.504283  0.391694  0.582738  0.152393  0.287706
7  0.174081  0.307924  0.586561  0.651703  0.463295  0.423682  0.329064
8  0.054052  0.136632  0.683219  0.585448  0.468530  0.214374  0.236304
9  0.054970  0.221971  0.291378  0.509513  0.437995  0.212444  0.479875

[10 rows x 401 columns]
```

Now, we can visualize our latent factors. We will create a scatterplot of the first and second latent factors in two-dimensional space using plt.scatter, after defining the x variable, xs, as the first factor and the y variable, ys, as the second factor. We label our x-axis using xlabel, our y-axis using ylabel, and our plot using title.

```
>>> xs = [w[0] for w in dtm_lsa]
>>> ys = [w[1] for w in dtm_lsa]
>>> plt.scatter(xs,ys)
<matplotlib.collections.PathCollection object at 0x000001F182CB77F0>
>>> xlabel('Latent Factor 1')
Text(0.5,0,'Latent Factor 1')
>>> ylabel('Latent Factor 2')
Text(0,0.5,'Latent Factor 2')
>>> title('Plot of First 2 Latent Factors')
Text(0.5,1,'Plot of First 2 Latent Factors')
>>> show()
```

The resulting scatterplot is displayed in Fig. 14.4.

Next, we will use the percentage of variance explained by retaining k factors to determine the number of latent factors to include in our LSA. Again, we use TruncatedSVD, although this time we will store the explained_variance_ratio_ attribute from the SVD as VarianceExplained. We will then plot this outcome as a scree plot, where we will look for an elbow point to choose the number of factors, k.

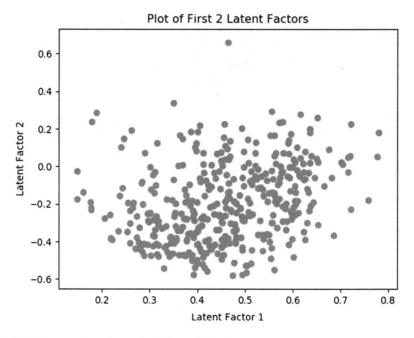

Fig. 14.4 First two latent factors in 25-factor LSA solution

```
>>> svd = TruncatedSVD(500, algorithm='randomized')
>>> svd.fit(dtm)
TruncatedSVD(algorithm='randomized', n_components=500, n_iter=5,
    random_state=None, tol=0.0)
>>> VarianceExplained=svd.explained_variance_ratio_
>>> plt.plot(VarianceExplained)
[<matplotlib.lines.Line2D object at 0x000002B23C898C18>]
>>> xlabel('k')
Text(0.5,0,'k')
>>> ylabel('Percent of Variance Explained')
Text(0,0.5,'Percent of Variance Explained')
>>> title('Scree Plot for up to 500 Latent Factors')
Text(0.5,1,'Scree Plot for up to 500 Latent Factors')
>>> show()
```

The resulting figure is shown in Fig. 14.5.
We can repeat the analysis using our *tfidf*-weighted DTM.

```
>>> lsa_tfidf = TruncatedSVD(25, algorithm='randomized')
>>> tfidf_lsa = lsa_tfidf.fit_transform(tfidf_dtm)
>>> tfidf_lsa = Normalizer(copy=False).fit_transform(tfidf_lsa)
```

Again, we can view the first 10 dimensions and the first and last 10 terms.

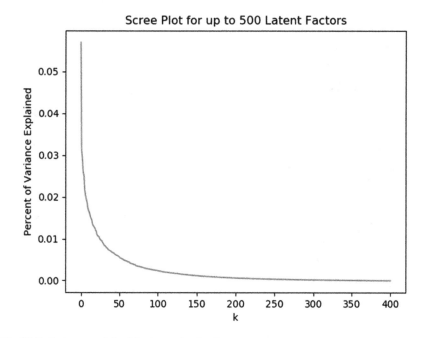

Fig. 14.5 Variance explained for increasing *k* values

```
>>> pd.DataFrame(lsa_tfidf.components_, columns=vectorizer.get_feature_names()).head(10)
```

	able	absolute	absolutely	absorb	accept	acceptable \
0	0.004532	0.003806	0.036714	0.001540	0.001892	0.000402
1	-0.000109	-0.000308	-0.019911	0.000587	0.000552	-0.000153
2	-0.011416	-0.011827	-0.012920	-0.002650	-0.002084	-0.001574
3	0.001706	-0.000245	0.002078	-0.001002	0.000167	0.000174
4	-0.006299	0.000210	-0.000423	-0.000069	-0.000995	-0.002080
5	0.006827	0.008616	0.005573	-0.001694	0.000160	-0.000086
6	0.007050	-0.003087	-0.055006	-0.000368	0.003015	0.000680
7	-0.003064	-0.016570	0.039658	-0.001583	-0.003136	0.000372
8	-0.001543	0.004540	-0.048210	-0.003343	-0.001629	0.000112
9	0.000628	0.013653	-0.043152	0.002352	0.002926	-0.000536

	accommodating	accommodation	account	achieved	...	years \
0	0.003376	0.000838	0.003654	0.000680	...	0.009959
1	0.000370	-0.000075	-0.002707	-0.000289	...	0.000703
2	-0.006854	-0.002223	-0.008213	-0.002345	...	-0.020144
3	-0.004910	0.000649	-0.000332	0.000040	...	0.017103
4	0.002401	-0.001766	-0.008210	-0.002383	...	0.010496
5	-0.001376	0.001034	-0.003302	-0.000389	...	0.024578
6	0.000418	0.001035	0.003157	0.001655	...	0.008065
7	-0.004491	-0.000652	0.001609	0.000216	...	-0.002677
8	-0.003113	-0.000225	0.004495	-0.001456	...	-0.023012
9	-0.001744	-0.002022	-0.003243	-0.001464	...	-0.032590

	yes	yesterday	yet	yiulyia	yoga	young	yuliya \
0	0.004061	0.005235	0.006019	0.001206	0.006747	0.007598	0.006310
1	-0.003922	-0.001345	-0.002894	-0.000754	0.000340	-0.005846	0.003899
2	-0.005922	-0.011012	-0.015730	-0.003295	-0.013280	-0.018710	-0.006131
3	0.003456	0.000727	-0.000160	0.000094	-0.002899	-0.001226	0.002607
4	-0.008747	-0.006797	-0.005336	-0.004458	-0.005088	-0.003650	0.003451
5	-0.000734	0.002281	-0.002739	-0.002551	0.002974	-0.002398	-0.003628
6	0.001322	0.009911	-0.006631	0.002644	0.001731	-0.016292	0.016666
7	0.008066	-0.001533	-0.012167	0.000104	-0.008068	0.003175	-0.010457
8	-0.007986	-0.008586	0.006275	-0.001241	0.003885	0.015251	-0.031549
9	0.002955	0.000858	0.011624	-0.002493	0.002930	0.019208	-0.001731

	zen	zero
0	0.008897	0.000714
1	0.014957	-0.000092
2	-0.002741	-0.002830
3	-0.028967	0.000324
4	0.030925	-0.002232
5	0.014698	0.002213
6	0.033013	0.002010
7	0.013009	-0.000939
8	0.022938	-0.001997
9	0.008754	-0.002966

[10 rows x 2305 columns]

We can again visualize our latent factors, this time based on the SVD using *tfidf* weighting. We create a scatterplot of the first and second latent factors in two-dimensional space using plt.scatter, after defining the x variable, xls, as the first factor and the y variable, yls, as the second factor. We label our x-axis using xlabel, our y-axis using ylabel, and our plot using title. The resulting plot is displayed in Fig. 14.6.

```
>>> xls = [q[0] for q in tfidf_lsa]
>>> yls = [q[1] for q in tfidf_lsa]
>>> plt.scatter(xls,yls)
<matplotlib.collections.PathCollection object at 0x0000029580E82A58>
>>> xlabel('Latent Factor 1')
Text(0.5,0,'Latent Factor 1')
>>> ylabel('Latent Factor 2')
Text(0,0.5,'Latent Factor 2')
>>> title('Plot of First 2 Latent Factors')
Text(0.5,1,'Plot of First 2 Latent Factors')
>>> show()
```

Again, we can evaluate how many dimensions to retain based on the percentage of variance explained. This time, we use the *tfidf*-weighted DTM. We will consider up to 25 dimensions and will view the singular vectors using print.

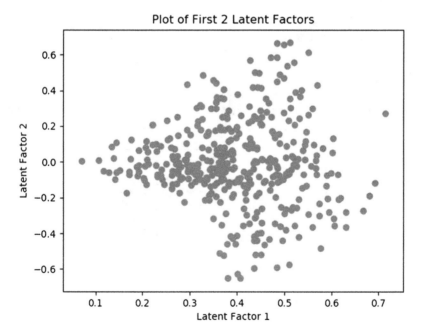

Fig. 14.6 First two latent factors in 25-factor LSA solution with *tfidf* weighting

```
>>> svd_tfidf = TruncatedSVD(25, algorithm='randomized')
>>> svd_tfidf.fit(tfidf_dtm)

TruncatedSVD(algorithm='randomized', n_components=25, n_iter=5,
    random_state=None, tol=0.0)
>>> print(svd_tfidf.singular_values_)

[4.45547573 2.95920593 2.72617011 2.51294425 2.26102596 2.19855807
 2.1328984  2.10353138 2.02640514 1.95992672 1.93506684 1.89246879
 1.85480958 1.83363862 1.79273831 1.77692858 1.7282294  1.72095243
 1.69238561 1.67376682 1.66262164 1.64495241 1.63905492 1.60095109
 1.59024176]
```

Visually, we can see that there is a big difference in the value of the first and second singular values, but all others exhibit less drastic differences. We can compute the variance explained by retaining k vectors and then visualize this result to determine how many singular vectors to keep in our LSA solution. The resulting scree plot appears in Fig. 14.7.

```
>>> VarianceExplained_tfidf=svd_tfidf.explained_variance_ratio_
>>> plt.plot(VarianceExplained_tfidf)

[<matplotlib.lines.Line2D object at 0x0000029580E717F0>]
>>> xlabel('k')

Text(0.5,0,'k')
>>> ylabel('Percent of Variance Explained')

Text(0,0.5,'Percent of Variance Explained')
>>> title('Scree Plot for up to 25 Latent Factors')

Text(0.5,1,'Scree Plot for up to 25 Latent Factors')
>>> show()
```

Based on the plot, there might be an elbow at $k = 4$. We can also view the values of the percent of variance explained.

```
>>> VarianceExplained_tfidf

array([0.0089098 , 0.02280897, 0.01812614, 0.01644031, 0.01329649,
    0.0125896 , 0.01179111, 0.01152468, 0.01067322, 0.01000166,
    0.00974257, 0.00931803, 0.00896053, 0.0087415 , 0.00837121,
    0.00822199, 0.00777894, 0.00770312, 0.00745876, 0.00728289,
    0.00719874, 0.00704623, 0.00699573, 0.00667463, 0.00658673])
```

As we suspected, there is a decrease of approximately 0.003. We can arrive at this result using the index values 3 and 4 of VarianceExplained_tfidf, which correspond to the singular values 4 and 5, because the index starts at 0.

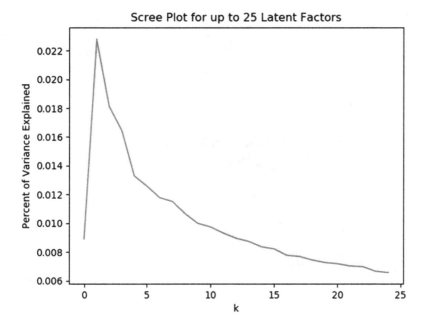

Fig. 14.7 Scree plot for up to 25 latent factors

```
>>> VarianceExplained_tfidf[3]-VarianceExplained_tfidf[4]

0.003143821356073063
```

Next, we can consider the top terms in the first dimension. We use the built-in components from our SVD. Then we sort the terms by weight, using .reverse to sort the terms in decreasing order by weight.

```
>>> sing_vecs = lsa_tfidf.components_[0]
>>> index = np.argsort(sing_vecs).tolist()
>>> index.reverse()
>>> terms = [vectorizer.get_feature_names()[weightIndex] for weightIndex in index[0:10]]
>>> weights = [sing_vecs[weightIndex] for weightIndex in index[0:10]]
>>> terms.reverse()
>>> weights.reverse()
```

Finally, we can plot the ten terms with the largest weight in the first dimension using plt.barh, which creates a barplot that is oriented horizontally. The resulting figure is displayed in Fig. 14.8.

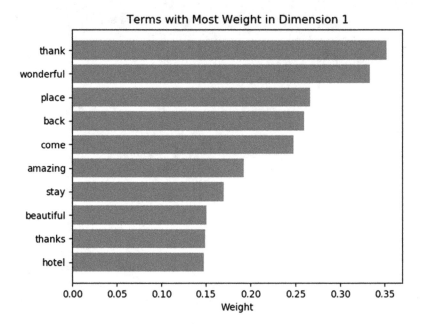

Fig. 14.8 Top 10 terms by weight for dimension 1

```
>>> plt.barh(terms, weights, align="center")

<BarContainer object of 10 artists>
>>> xlabel('Weight')

Text(0.5,0,'Weight')
>>> title('Component 1 Terms with Most Weight')

Text(0.5,1,'Component 1 Terms with Most Weight')
>>> show()

>>> plt.barh(terms, weights, align="center")

<BarContainer object of 10 artists>
>>> xlabel('Weight')

Text(0.5,0,'Weight')
>>> title('Terms with Most Weight in Dimension 1')

Text(0.5,1,'Terms with Most Weight in Dimension 1')
>>> show()
```

As shown, the top terms are overwhelmingly positive terms, such as *thank*, *thanks*, *amazing*, *beautiful*, and *wonderful*. We can automate this process for the first four dimensions by using a loop.

```
>>> result=[]

>>> for i in range(0,4):
    sing_vecs = lsa_tfidf.components_[i]
    index = np.argsort(sing_vecs).tolist()
    index.reverse()
    terms = [vectorizer.get_feature_names()[weightIndex] for weightIndex in index[0:10]]
    weights = [sing_vecs[weightIndex] for weightIndex in index[0:10]]
    terms.reverse()
    weights.reverse()
    temp = pd.DataFrame(columns=('terms','weights'))
    temp['terms'] = terms
    temp['weights'] = weights
    result.append(temp)
```

First, we can view the top 10 terms for the first dimension, indexed by 0, to view the format of the results of our loop, which store the dataframe results as a list called "result."

```
>>> result[0]

          terms        weights
0         hotel        0.147558
1         thanks       0.149188
2         beautiful    0.150366
3         stay         0.170099
4         amazing      0.192657
5         come         0.247706
6         back         0.259533
7         place        0.266045
8         wonderful    0.333778
9         thank        0.352487
```

Next, we can use subplots to show the four plots for the four dimensions on the same plot. Again, we will use a loop to create the subplots efficiently and then plot the main figure. The top words for the dimensions are displayed in Fig. 14.9.

```
>>> fig = plt.figure()
>>> fig.subplots_adjust(hspace=.5, wspace=.5)
>>> for i in range(0, 4):
    ax = fig.add_subplot(2, 2, i+1)
    ax.barh(result[i]['terms'],result[i]['weights'], align="center")
    ax.set_title('Dimension %d' % (i))

<BarContainer object of 10 artists>
Text(0.5,1,'Dimension 0')
<BarContainer object of 10 artists>
Text(0.5,1,'Dimension 1')
<BarContainer object of 10 artists>
Text(0.5,1,'Dimension 2')
<BarContainer object of 10 artists>
Text(0.5,1,'Dimension 3')
>>> plt.show()
```

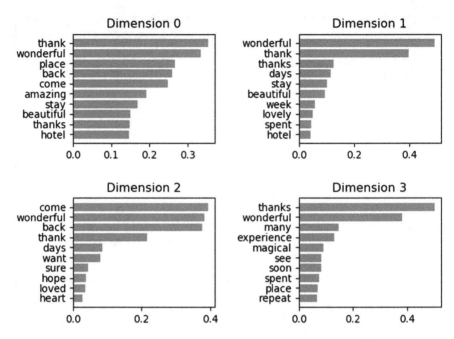

Fig. 14.9 Top 10 terms in first four LSA dimensions

In viewing the weights of the terms, we see that several terms factor strongly into more than one of the dimensions.

Key Takeaways
- LSA is implemented in the open-source Python software in a step-by-step analysis.
- The modeling capabilities of Python make it an ideal tool for analyzing data from sources that generate large corpora, such as social media, online reviews, and open-ended questions in surveys.

Acknowledgment   The authors thank Jorge Fresneda Fernandez, Assistant Professor of Marketing at the Martin Tuchman School of Management, New Jersey Institute of Technology for contributing this chapter to the book.

Further Reading

To learn more about the open-source Python software, visit https://wiki.python.org/moin/BeginnersGuide/Download

Chapter 15
Learning-Based Sentiment Analysis Using RapidMiner

Abstract This chapter provides a step-by-step sentiment analysis in RapidMiner using classification analysis. After being introduced to the RapidMiner software, the reader learns to build a process map-based analysis to classify Amazon reviews by sentiment. Two machine learning methods, k-nearest neighbor and naïve Bayes, are demonstrated and assessed for predictive performance.

Keywords RapidMiner · Sentiment analysis · Categorization · Classification analysis · k-nearest neighbor · Naïve Bayes · Online consumer reviews

15.1 Introduction

RapidMiner is a process map-based software program for data science. In this example, we use RapidMiner Studio, which can be downloaded at: https://my.rapidminer.com/nexus/account/index.html#downloads.[1] We will also use the Text Processing extension. Information about the extension can be found at: https://marketplace.rapidminer.com/UpdateServer/faces/product_details.xhtml?productId=rmx_text.

Online consumer reviews (OCR) are an important source of information for companies to monitor customer satisfaction. The ability to predict the sentiment of an online review can help inform marketing strategies. One method that can be used to build a predictive model of consumer sentiment is learning-based sentiment analysis. In this step-by-step example using RapidMiner, we perform a sentiment analysis using classification methods. We use 1,000 sentiment labeled Amazon reviews, with sentiment label 0 for negative reviews and sentiment label 1 for positive reviews.[2] Since the sentiment classification of the reviews is known, we use two classification methods, naïve Bayes and k-nearest neighbor, to build predictive

---

[1] For more information about getting started with RapidMiner Studio, visit https://docs.rapidminer.com/latest/studio/

[2] The data can be downloaded from https://archive.ics.uci.edu/ml/machine-learning-databases/00331/. The description of the dataset can be found at https://archive.ics.uci.edu/ml/datasets/Sentiment+Labelled+Sentences#

© Springer Nature Switzerland AG 2019
M. Anandarajan et al., *Practical Text Analytics*, Advances in Analytics and Data Science 2, https://doi.org/10.1007/978-3-319-95663-3_15

models. Once the models are built, we can assess their accuracy by viewing a contingency table of the predicted versus actual classifications and comparing the models based on their predictive performance.

15.2   Getting Started in RapidMiner

To begin, we download RapidMiner Studio. Once the download is complete and RapidMiner is installed, we run RapidMiner. At the RapidMiner welcome screen, we choose "New Process" and "Blank" to create a blank process template, shown in Fig. 15.1.

There are three views available in RapidMiner Studio: Design, Results and Auto Model. The Design view will be used to create our analysis process. In the Design view, we have tabs on the left-hand side named "Repository" and "Operators." Data can be imported in the Repository tab, but we will focus on the Operators tab. Operators are the building blocks of the process map, which can be used to import, transform, and analyze data (Hofmann and Klinkenberg 2013). The main folders containing the operators are the following: Data Access, Blending, Cleansing, Modeling, Scoring, Validation, Utility and Extensions. The Extensions folder contains add-on extensions.

On the right-hand side, the tabs are named "Parameters" and "Help." As we set up our process map in the main Process panel in the center of the Design view layout, if there are parameters that can be changed, we will update them in the Parameters tab. The main Parameters options include the ability to choose a random

Fig. 15.1 RapidMiner welcome screen

Fig. 15.2 Main RapidMiner view

Fig. 15.3 RapidMiner view with operations, process, and parameters panels

seed. The center Process panel will be where we create the process map for our analysis. We can drag individual operators to the Process panel to add them to our analysis. On either end of the Process panel, there are two ports: inp and res. The inp port on the left stands for input, and the res port on the right stands for results. To view the results of the process, we must connect the last operator in the process to the res port. The main screen in RapidMiner is shown in Fig. 15.2.

Since we will not use the Repository and Help panels, we can remove them by clicking the × next to these panels (Fig. 15.3).

Fig. 15.4 Marketplace extension drop-down menu

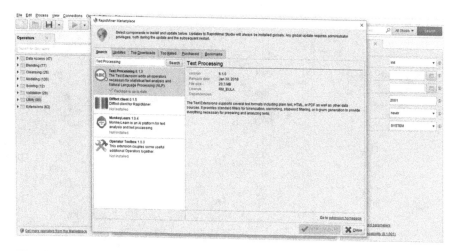

Fig. 15.5 Text Processing extension

For our analysis, we will use the Text Processing extension, which we can add in RapidMiner Studio. We can add this extension by using the top menu bar. Navigate to the "Extensions" on the top menu bar and choose "Marketplace" from the drop-down menu (Fig. 15.4).

A pop-window with a Search tab will appear. Here, we can search for "Text Processing." If the Text Processing extension is not already installed, we can click the "Install" button in the bottom right-hand corner of the pop-up screen to add the extension. Once the installation is completed, we will be able to view the Text Processing extension in a subfolder of the extensions folder in the Operators tab (Fig. 15.5).

Now that we have familiarized ourselves with RapidMiner and added the necessary extension, we are ready to import and prepare our text documents for
analysis.

15.3 Text Data Import

We use the Operators panel to select the Process nodes for text preprocessing. The
data for analysis are in the form of a text file, which we can read as a csv file by
changing the delimiter. We highlight and drag the "Read CSV" operator located in
the Data Access file in the Operators tab to our Process panel (Fig. 15.6).

When the Read CSV node is highlighted, we edit the node in the Parameters
panel on the right-hand side of the Studio window. Here, we can use the Input
Configuration Wizard button to import our data more easily. First, choose the file
from the location where the text file is saved. Be sure to change the type of file to
"All Files" because we are importing a text file rather than a csv file. Then, choose
the amazon_cells_labelled.txt file and choose "Next." (Fig. 15.7).

Then, we specify how the file should be imported. We choose not to skip comments or trim lines. We choose tab as the delimiter, rather than comma, and uncheck
the "Use Quotes" box. Then, click "Next." (Fig. 15.8).

Next, we use the drop-down next to the first row to indicate that the first row does
not contain columns names. Then, click "Next." (Fig. 15.9)

Finally, we change the attribute data type and names. In RapidMiner, data variables are known as attributes. First, we can use the drop-downs to indicate that att1
is text and att2 is nominal, because the 0 s and 1 s stand for negative and positive
sentiments. On this screen, we can also change the column names, which automati-

Fig. 15.6 Read CSV node

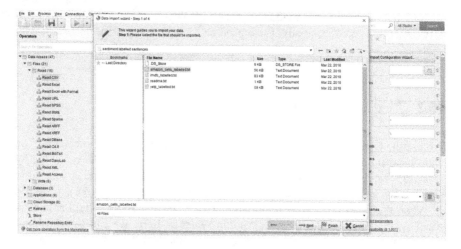

Fig. 15.7 Data import wizard

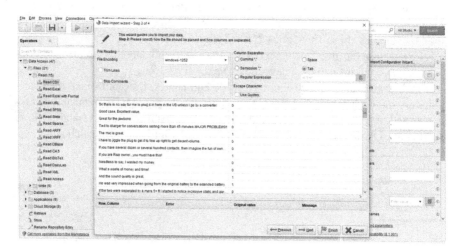

Fig. 15.8 Tab delimited text

cally are named att1 and att2, for attribute 1 and attribute 2. Here, we can rename the columns as "text" and "sentiment." For our analysis, the text is an attribute, but the sentiment is a label. We use the drop-down to change sentiment to a label. Finally, we can click "Finish." (Fig. 15.10)

By connecting the output port on the Read CSV operator to the res port on the right-hand side of the Process panel, we can set up our process to run (Fig. 15.11).

To run the process, we can then choose the blue arrow or use a keyboard shortcut, F11. Once the process is run, the Results tab is automatically displayed with a preview of the data (Fig. 15.12).

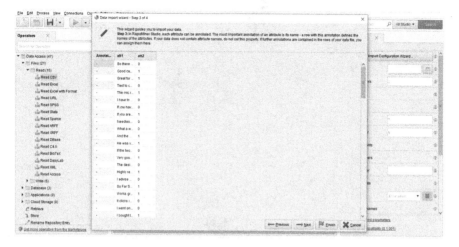

Fig. 15.9 Row name designations

Fig. 15.10 Data type designations

15.4   Text Preparation and Preprocessing

Next, we return to the Design view to process our data. To do this, we navigate to the Text Processing folder, which is a subfolder of the Extensions folder. In the Text Processing folder, we highlight and drag the "Process Documents from Data" operator. By dragging it onto the connection between the Set Role node and the res port, the connections will automatically be updated. In the Parameters panel, we can choose the weighting schema used. The default weighting is *tfidf*. In this example, we use term frequency, because we will be classifying sentiment polarity using k-nearest neighbor (kNN) and naïve Bayes (NB) (Fig. 15.13).

Fig. 15.11 Read CSV connection to Results port

Fig. 15.12 Read CSV results

Note that there is a yellow triangle on the Process Documents operator, indicating that we have a warning. We can right-click on this operator and choose "Show warnings." Based on the warning, we now have a new screen for the "Process Documents from Data" operator. On this new screen, we will add our preprocessing operators that will be completed by the Process Documents from Data operator (Fig. 15.14)

Next, we drag the Tokenize operator, which is located in the Tokenize subfolder of the Text Processing folder in the Operator panel, to our Process panel and connect it to doc (document) ports on either side. In the Parameters panel, we can specify the tokenization mode, which defaults to non-letters. For our analysis, we will use the default mode (Fig. 15.15).

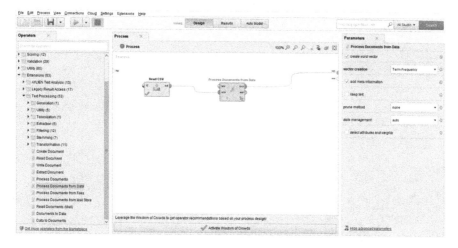

Fig. 15.13   Process Documents from Data operator

Fig. 15.14   Process Documents from Data operator warning

Again, we can click the "Run" button or use the keyboard shortcut to view the results of the current process map (Fig. 15.16).

As we can see, a document-term matrix (DTM) is created from our review documents, which includes sentiment as the label attribute. In viewing the terms, we see that additional preprocessing must be completed, including converting the text to lower case. Returning to the Design view, we use the "Transform Cases" operator in the "Processing Text" subfolder called "Transformation" to achieve this goal. We click and drag Transform Cases onto the line between the Tokenize and final doc port. From the Parameters panel, we see that the default transformation is to lower case (Fig. 15.17).

Fig. 15.15 Tokenize operator

Fig. 15.16 Tokenize operator results

Next, we can remove stop words. As in other software programs with text processing capabilities, RapidMiner has built-in stop word lists, which can be accessed in the "Filtering" subfolder. Here, we can choose the Filter Stop words (English) operator and drag it after our Transform Cases node. As shown, this operator does not have any parameters (Fig. 15.18).

Next, in the Filtering folder, we add the "Filter Tokens (by Length)" operator after the Filter Stop words (English) node to set a minimum and maximum character

Fig. 15.17 Transform Cases operator

Fig. 15.18 Filter Stop words (English) operator

length for our tokens. Based on the Parameters panel, the default minimum is 4 and the default maximum is 25. In this analysis, we do not change the defaults (Fig. 15.19).

For our final preprocessing step, we use the Porter stemming algorithm to stem the tokens. The Stemming subfolder contains many stemming options, including Snowball, Porter, and Lovins. We choose the "Stem (Porter)" operator and drag it onto the connector after the Filter Stop words (English) node (Fig. 15.20).

Again, we see that this operator does not have any parameters. Having completed all of the preparation and preprocessing steps, we can view the resulting DTM by running the process. As shown, there are 1,000 documents and 1,174 terms in our DTM (Fig. 15.21).

Fig. 15.19 Filter Tokens (by Length) operator

Fig. 15.20 Stem (Porter) operator

To explore the term frequency, we can return to the beginning nodes and add an additional connection from the wor (word) port on the Process Documents from Data operator to the second res port (Fig. 15.22).

Now, if we run the process, we can expect to have two result tabs: one tab with the DTM and one tab with a term list and frequency information. Here, we have columns displaying the term frequency and document frequency values for all of the terms in our DTM (Fig. 15.23).

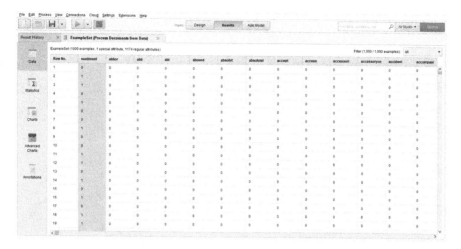

Fig. 15.21 Process Documents operator results

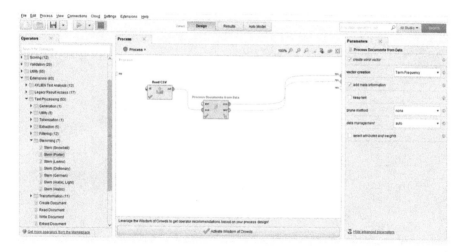

Fig. 15.22 Term and document frequency node connections

By clicking on the column names of the WordList, we can sort by specific columns. First, we sort the Total Occurrences column in descending order of frequency to view the most frequent terms. As shown, *phone* is the most frequently used word in the corpus, with a term frequency of 178, followed by *work* (112), *great* (99), *good* (78), and *product* (56) (Fig. 15.24).

Next, we sort the Document Occurrences column in descending order of frequency to view the terms with the highest document frequency. As shown, the terms with the highest term frequency also have the highest document frequency. *Phone* has the highest document frequency (167), followed by *work* (108), *great* (97), *good* (75), and *product* (56) (Fig. 15.25).

Fig. 15.23 Term and document frequency results

Fig. 15.24 Term occurrences sorted in descending order

15.5 Text Classification Sentiment Analysis

To begin, we need to add a Split Validation operator to the main process screen to split the data into training and testing. The Split Validation operator is in the Validation folder. We drag this operator to a position after the Process Documents from Data node and connect the exa (example set) port from the Process Documents node to the tra (training) port on the new Validation node. We can also disconnect the connection between wor and res. Now, we connect the mod (model) port and one of the ave (average) ports on the Validation operator to the res ports. We use the default split criteria, which is to put 70% of the observations in the training data and

Fig. 15.25 Document Occurrences sorted in descending order

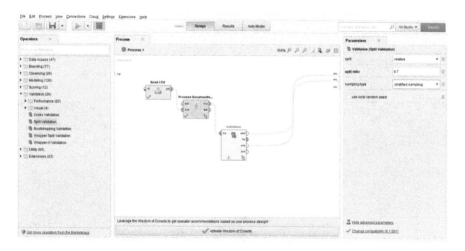

Fig. 15.26 Validation operator

30% in the testing data. In this case, the auto-sampling feature is a stratified sample, with proportionate class representation in the training and testing data (Fig. 15.26).

Now, if we click the Process drop-down menu, we can go to the Validation node view to set up the training and testing configuration (Fig. 15.27).

On the left-hand side of the Validation pane, we see the training pane, and the testing pane is on the right side. First, we need to specify the model for training. We will use a kNN operator node, which we drag from the "Lazy" subfolder in the "Modeling" folder. We connect the two tra (training) and mod (model) ports. On the Parameters panel, we can update the number of nearest neighbors, k. For our example, we will choose $k = 5$ (Fig. 15.28).

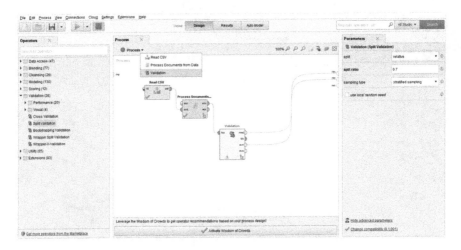

Fig. 15.27 Validation node view drop-down

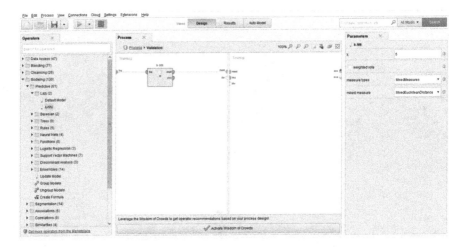

Fig. 15.28 kNN operator

Next, we need to set up the training pane. In this pane, we first add an "Apply Model" operator from the "Scoring" folder and connect the mod ports and the test (testing) and unl (unlabeled) ports on the left-hand side of the Apply Model node. Then, to the right of the Apply Model node, we add a "Performance (Classification)" operator from the "Predictive" subfolder in the "Validation" folder. We connect the two lab ports and finally connect the per port on the lef (left) to the ave (average) port on the right. In the Parameters pane of the Performance node, we can choose additional performance measures to display in the results of the analysis (Fig. 15.29).

Fig. 15.29 Apply model and performance operators in validation view

accuracy: 63.00%

	true 0	true 1	class precision
pred 0	142	103	57.96%
pred 1	8	47	85.45%
class recall	94.67%	31.33%	

Fig. 15.30 kNN results contingency table

Next, we can run the model to view the confusion matrix of classifications of the testing data. As shown in the upper right of the results, the accuracy of the kNN model is 63%. The recall and precision values for the sentiments are also displayed in the contingency table (Fig. 15.30).

To complete the naïve Bayes analysis, we return to the Validation pane view in the Design view to replace the kNN node. We can remove the kNN node by right clicking on it and choosing "Delete." (Fig. 15.31).

Once the kNN node is removed, we can add the "naïve Bayes" operator, which is located in the "Bayesian" subfolder of the "Modeling" folder. We drag this operator to the training pane and replace the connections between the tra and mod nodes (Fig. 15.32).

Finally, we can run the NB model to view the contingency table based on the predicted and actual classes for the reviews. As shown in the Results pane, the NB model outperforms the kNN model, with 73% accuracy. The class-level precision and recall values are also displayed in the confusion matrix (Fig. 15.33).

Fig. 15.31 Remove kNN operator

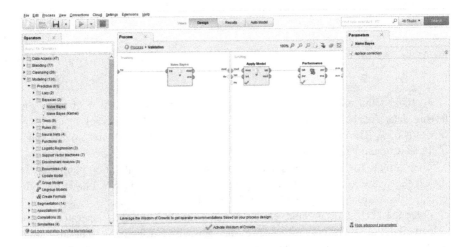

Fig. 15.32 Naïve Bayes operator

accuracy: 73.33%

	true 0	true 1	class precision
pred. 0	88	18	83.02%
pred. 1	62	132	68.04%
class recall	58.67%	88.00%	

Fig. 15.33 Naïve Bayes results contingency table

> **Key Takeaways**
> - A step-by-step classification analysis using the process map-based RapidMiner Studio software is demonstrated.
> - Learning-based classification methods, including naïve Bayes and k-nearest neighbor, are demonstrated.
> - The RapidMiner Studio provides an interactive user interface for noncoders to perform text analytics.

Reference

Hofmann, M., & Klinkenberg, R. (Eds.). (2013). *RapidMiner: Data mining use cases and business analytics applications*. Boca Raton: CRC Press.

Further Reading

For more about RapidMiner, see Hofmann and Klinkenberg (2013).

Chapter 16
SAS Visual Text Analytics

Abstract This chapter presents a step-by-step visualization analysis of over 4,000 health news tweets using SAS Visual Text Analytics (VTA). SAS VTA is a commercial software program that uses a pipeline, or process-based, approach to the analysis of text. This chapter demonstrates the creation of visualizations including tree maps, line charts, pie charts, and word clouds using the software.

Keywords SAS Visual Text Analytics (VTA) · Twitter · Health news · Text data visualization · Social media

16.1 Introduction

This chapter covers the application of the SAS tool Visual Text Analytics (VTA). VTA combines traditional natural language tools and linguistic rules with other more advanced tools, such as machine learning, to reveal and visualize the most relevant insights contained in textual data. VTA supports four different types of models that address a broad range of text analysis needs: categories, contextual extraction, sentiment, and text topics.[1] VTA integrates with open-source languages such as R or Python and allows users to incorporate SAS Text Miner capabilities into the analysis. Additionally, VTA projects can be integrated into SAS Platform, which facilitates collaborations and exchanges among users.

VTA adds some unique features to text analysis. One of the most interesting features is the ability to combine machine learning and linguistic rules. This feature allows users to incorporate elements such as slang, irony, and sarcasm with which regular machine-learning approaches struggle. VTA provides tools to handle context-specific content, such as expressions, names, dates, part-of-speech tags, currency, and measurements. VTA offers support for the text analysis of 30 different languages including default stop word lists for each of those languages.

---

[1] https://rpclab02023.sas.com/epmenu/pdf/Telecommunications_Discovery_VTA_Script.pdf

© Springer Nature Switzerland AG 2019
M. Anandarajan et al., *Practical Text Analytics*, Advances in Analytics and Data Science 2, https://doi.org/10.1007/978-3-319-95663-3_16

VTA is more user-friendly than other approaches and easy to implement. Users can take advantage of its visual programming flow (called a "pipeline" in VTA), which includes nodes that can be controlled independently and provides flexibility to the entire analysis.

On the negative side, SAS VTA is proprietary software, and therefore users must pay a fee to access the tool after the trial period. Additionally, VTA is constrained to the options provided by SAS and lacks the flexibility (which also translates into more complexity) of open-source options, in which the user can "craft" the analysis to meet the very specific requirements.

In this chapter, we will run VTA over a dataset that includes 4,141 health news tweets from the Twitter account of the *Los Angeles Times* to illustrate our analysis— these tweets were posted from December 2011 to March 2015.[2] Prior to the implementation of the analysis itself, the dataset was manipulated to remove repeated instances—some tweets were posted twice or even three times and the links provided to read the full news report when those links were available in the tweet itself. "Artificially" redundant data—such as the same tweet repeated several times—can have a significant impact on the results, regardless of the text analysis tool employed. The analyses will be run using the SAS Visual Text Analytics version 8.2 on the cloud platform SAS Viya. Due to space constraints, we will address only a few of the functionalities available through VTA.

16.2   Getting Started

After logging into the SAS Viya account, users can access the most relevant options related to the data management, data analysis, and environment management through the Welcome page, as shown in Fig. 16.1. This option includes data management, data preparation, data exploration and visualization, model building, model management, decision management, workflow management, and environment management.

The data to be analyzed—in our case the 4,141 health news tweets—can be loaded by selecting the *Manage Data* icon on the Welcome page. SAS VTA makes the data importation very easy. The Manage Data option allows users to handle the already available datasets and to import local data files from the user's system. After selecting the option *Import*, users can either select the location of the data file or drag the data file themselves into the drop area as shown in Fig. 16.2.

The data are imported from their location on the computer, and a confirmation message is generated after the dataset is correctly loaded onto the platform.[3]

---

[2] Data available on https://archive.ics.uci.edu/ml/datasets/Health+News+in+Twitter#

[3] Note that the data were imported in a tab-delimited, .txt format and using UTF-8 encoding. This file is called LATIMESHEALTH4, as shown in the screenshots.

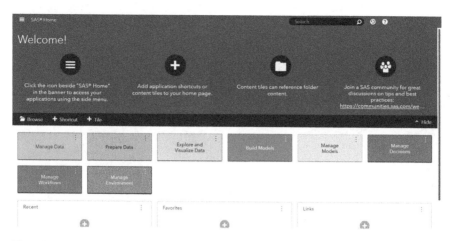

Fig. 16.1 SAS Viya Welcome page

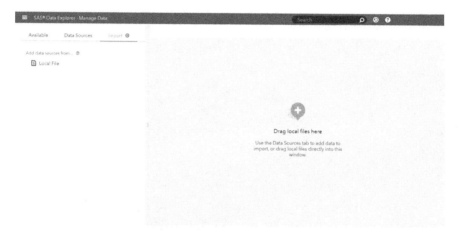

Fig. 16.2 Data import option for data management

The data are now available in the system. To confirm that availability, users should be able to see a short description of the dataset through the *Available* option as shown in Fig. 16.3. The right-hand side of the screen of *Details* shows that the file contains 4,141 rows and only one column, *CleanText*, corresponding to the text of the 4,141 tweets.

Users can also confirm that the dataset was correctly loaded by inspecting a few rows through the *Sample Data* tag within the *Available* option. The default option is to inspect 100 rows, although users can change the number of rows they want to review. Figure 16.4 shows the first 13 rows of the dataset.

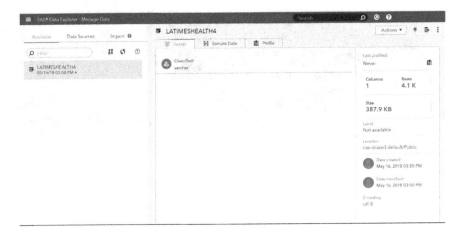

Fig. 16.3 The *Available* option shows the dataset loaded

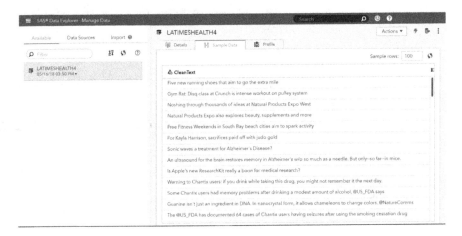

Fig. 16.4 First 13 rows in the dataset

16.3   Analysis

Once the dataset is available, users can start building the analysis pipeline. The Welcome page allows the creation of the analysis pipeline through the *Build Model* tag. (Note that the Welcome page can be accessed anytime through the top left-hand corner *Show Applications Menu* tag). We should select the option *New Project* in order to begin the analysis. A pop-up screen will allow us to enter three important elements: the name of the new project, the type of project (such as *Data Mining and Machine Learning*, *Forecasting*, or *Text Analytics*), the data source, and the language of the new project. After naming the new project, we select the *Text Analytics* option, the Tweets dataset, and the *English* option for language. These four steps allow us to create the new project.

Fig. 16.5 Assigning the text variable role

Fig. 16.6 Customizing the pipeline

VTA requires assigning a variable to the text role, which we can do through the *Assign Variable Role* option. By clicking this option, a new pop-up screen will show a warning message indicating that no text variable has been selected, as shown in Fig. 16.5.

For this analysis, we assign the only variable available (named "CleanText") to *text variable*. As explained in the Introduction section, "CleanText" contains the 4,141 unique tweets with the links to the full news reports removed (Note that the warning message has now disappeared, as shown in Fig. 16.5). At this point, the process should continue by adapting the pipeline to our analysis requirements. A pipeline is the way in which SAS VTA represents the sequence of tasks performed in the analysis. Each of these independent tasks constitutes a node in the pipeline. The tag *Pipelines* is located on the top left-hand side of the screen, as Fig. 16.6 illustrates.

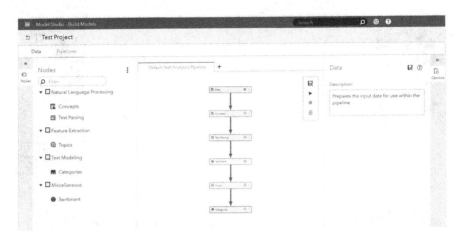

Fig. 16.7 Default pipeline available on VTA

The default pipeline shows six independent elements: *Data*, *Concepts*, *Text Parsing*, *Sentiment*, *Topics*, and *Categories*. Pipeline nodes can be somewhat modified prior to the analysis, and users can add or remove nodes from the default pipeline. Figure 16.7 shows the default pipeline.

The first node, the *Data* node, prepares the textual data to be used in the pipeline. The *Concepts* node extracts specific information from the textual data, as the analyst needs it, because it might be relevant within some specific context. The user can decide whether to implement the analysis with the predefined concepts. For instance, these predefined concepts can identify names, places, persons, dates, or organizations. The predefined concepts appear in Fig. 16.8.

The following node, *Text Parsing*, allows the user to get the text ready for term analysis. The options that can be adjusted for the analysis include the minimum number of documents in which a term must appear to be included in the analysis; the use of either a start list, a list of terms that can be included in the analysis, or a stop list, a list of terms that cannot be included in the analysis; and the use of a synonym list. (Note that a start list and a stop list cannot be employed simultaneously.) These options appear in Fig. 16.9. Both *Concepts* and *Text Parsing* are the default natural language processing nodes.

The next node is *Sentiment*. The sentiment model identifies the attitudes included in the textual data and can generate a score for feature-level sentiments. By adding analysis nodes following the *Sentiment* node, users can develop a document-level analysis of sentiments.

Following the workflow elements of the pipeline, the next default node corresponds to *Topics*. The basic function of this node is to assign the documents analyzed to topics. Users can select to implement automatic topic detection or to specify the number of topics in advance. If the latter option is selected, the analyst must

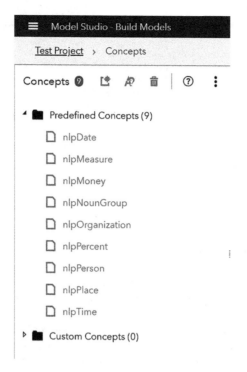

Fig. 16.8 Predefined concepts available

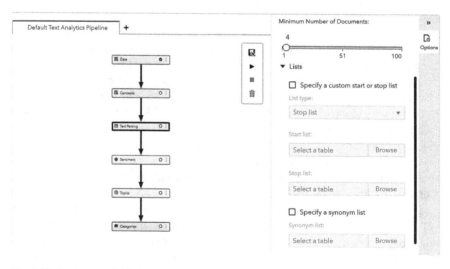

Fig. 16.9 Options available for text parsing

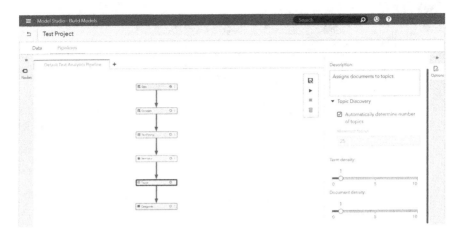

Fig. 16.10 Topic node options

enter the maximum number of topics. Otherwise, automatic detection must be selected. The topic selection is implemented through unsupervised machine learning methods.

Users can also customize both the *term density* and the *document density*. The term density refers to the cutoff value needed to include a term in the topic, relative to the absolute value of the weight of that term. By weight, VTA means the importance of one term relative to the entire population of terms in the data. The document density is similar but differs in that it assigns the cutoff value for the number of documents assigned to a topic. Term density and document density can take values between 0 and 10. The closer to 0, the more densely populated the topics will be. In contrast, increasing the values closer to 10 will create very selective, less densely populated topic. Figure 16.10 displays these options. *Sentiments* and *Topics* are the two default nodes for feature extraction.

The last default node is *Categories*. This node classifies documents by subject. Users can either allow the software to generate category rules or provide rules for the category variables. In VTA, categories are a higher-level classification element than topics. This hierarchy implies that topics can be "promoted" to categories, but categories cannot be "promoted" to topics. The classification of documents is based on linguistic rules instead of the weighting of terms, as is the case in topics. (Note that documents can be categorized as belonging to more than one category.) *Categories* is the default node for text modeling.

In this example, (1) we included the predefined concepts in the *Concepts* node; (2) we used the predefined English stop list that is applied automatically and adjusted the minimum number of documents to 10 within the *Text Parsing* node; (3) within the *Topics* node, we selected the "Automatically determine number of topics" option and adjusted the term and document density to 3 to obtain a lower density in the results than with the default value of 1; and (4) we selected "Automatically generate categories and rules" under the *Categories* node. (Note that due to the nature of the data, health news tweets, which are very short pieces of

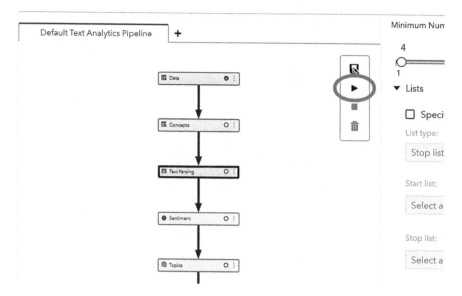

Fig. 16.11   Run pipeline icon to implement the analysis

text, we did not select the option "Specify a sentiment model" within the *Sentiment* node.) After the selection of these settings, the analysis is ready to be implemented. We can start the analysis by either clicking on the "Run pipeline" icon on the top right-hand side of the pipeline area (shown in Fig. 16.11) or right clicking on one node and selecting "Run."

VTA runs the analysis sequentially, following the pipeline flow, and generates a warning message if it encounters an issue while performing the specified tasks within each node. (Note that in Fig. 16.12, all of the check marks appear in green. If VTA encounters a problem, it displays a red mark on the specific node that generated the problem, and the analysis stops at that specific node. If such a problem occurs, users should modify the settings of the specific node that created the problem in order to complete the entire analysis.)

By right clicking on *Concepts* and selecting "Open," we can access the predefined concepts referred to previously. As an example, by selecting the "nlpPerson" predefined concept under matched documents, we can identify names such as Angelina Jolie, Ray Romano, and James Gandolfini, because they were mentioned in the tweets included in the dataset. This example appears in Fig. 16.13.

Similarly, selecting "nlpOrganization" allows us to identify organizations mentioned in the tweets such as the Pentagon, the Supreme Court, and the USDA. Clicking "Close" returns us to the pipeline screen.

By right clicking on *Text Parsing* and selecting "Open," we can access the list of terms that were kept and the list of terms that were dropped from the analysis. The list of dropped terms results from the application of the default stop word list over the dataset. Figure 16.14 shows the first six elements of this list on the right. Words such as "be," "to," "the," "a," "in," or "of" are very common, but they do not

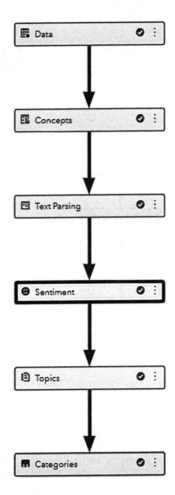

Fig. 16.12 The pipeline tasks completed

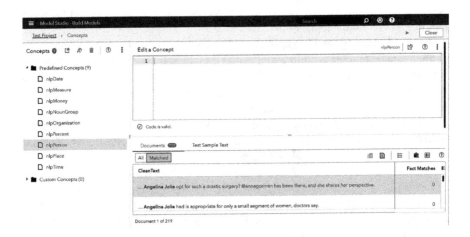

Fig. 16.13 Predefined concepts *nlpPerson*

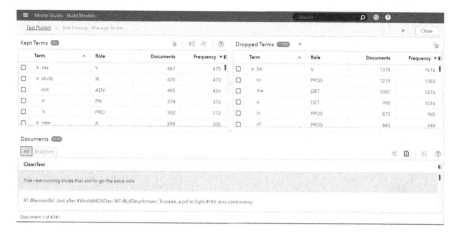

Fig. 16.14 List of kept and dropped terms

Fig. 16.15 Results of topic analysis

carry relevant meaning and therefore were dropped from the analysis. By clicking "Close" we can go back to the pipeline screen.

By right clicking on *Topics* and selecting "Open," we can access the automatic topic selection requested. The results show 12 different topics and 657 terms included in the analysis, as illustrated in Fig. 16.15.

The majority of the topics can be easily labeled. For instance, the first one includes terms such as "heart," "risk," "disease," "heart disease," and "attack." Users can easily infer that this topic refers to heart diseases/heart attacks. To delve even more into the identification of each topic, users can inspect the list of relevant terms associated with it by clicking the "Matched" option of the term list on the right. They can also review how these terms were used in the documents by selecting the "Matched" option of the document section at the bottom. These options are illustrated in Fig. 16.16. (Note that the relevancy of each term and each document

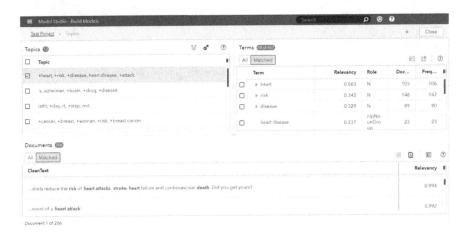

Fig. 16.16 Most relevant terms associated with topics and their use in documents

in the topic also appear in this node). Clicking "Close" returns us to the pipeline screen to continue checking the results of the analysis.

The *Categories* node can be opened by right clicking on it and selecting "Open." The automatic generation of categories and rules did not create any categories, so we defined three custom categories based on the inspection of the dataset. These three custom categories are: (1) "disease/condition," (2) "research," and (3) "legal." By right clicking on *All Categories*, "Add new category," we can add new customized categories.

After choosing to add a new category, users are prompted to enter the name of the new category. We named all three categories ("disease/condition," "research," and "legal") and provided a definition for each one, including the terms that we wanted to use to define those categories. In the case of disease/condition, the words selected were "heart," "risk," "disease," "attack," "Alzheimer," "brain," "cancer," and "breast." For research, the words chosen were "new," "research," "drug," "health," "study," and "find." Finally, for legal, the list of words was "healthcare," "law," "supreme", and "court."

We indicate the definition of the categories by entering a simple code (technically speaking, a category rule) in the "Edit a Category" pane, as shown in Fig. 16.17. Each category rule includes two major elements: operators (such as Boolean operators) and arguments (such as the words selected for each category). As Fig. 16.17 illustrates, operators appear in blue color, while arguments are in purple. The code (or rule) should be validated (by clicking on the "Validate Rule" icon on the top right) so that no mistakes are made in entering the definition of the category.

The category rules are a very important element of the analysis. Therefore, users should ensure that their definition is correct. These category rules should be tested prior to applying the changes to the entire analysis. To do so, we can select an

Fig. 16.17 Category definition: code and validation

Fig. 16.18 Category rule validated over a sample document

individual document containing words that were specified as part of the category rule and test whether the system identifies the matched items. We selected the disease/condition category, and we also selected a document containing two words included in this category rule, "disease" and "Alzheimer," within the *Documents* tab. Once the document was selected within the *Documents* pane, we clicked on the "Paste to Test Sample Text" icon and moved to the *Test Sample Text* pane, where we ran the test by clicking on "Test text." As Fig. 16.18 illustrates, the system correctly identified the two words "disease" and "Alzheimer," and therefore the category rule was validated.

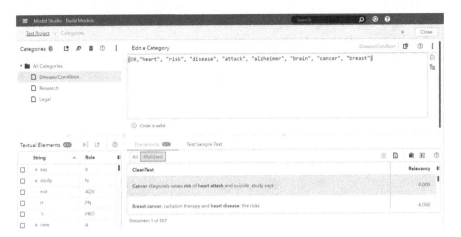

Fig. 16.19 Example of matched documents in the disease/condition category

Once the category rules were provided and validated, we ran the analysis again. Opening the category node allows us to check the results of applying the category definition. As Fig. 16.19 illustrates, users can select one of the three categories defined and inspect the matched documents. In this case, we reviewed the disease/condition category. The relevancy of the document to the category also appears on the right-hand side of the "Matched" pane.

At this point, we closed the category node and returned to the pipeline screen.

Additional explorations and visualizations are available on VTA if users decide to export the categorization model created. First, users must save the categorization analysis results by right clicking on the *Categories* node and selecting "Save data table." Under *Data Sources*, users need to expand *cas-shared-default* and select their own CASUSER(XXXX@XXXX) library to save the data table. This action allows users to start the advanced explorations and visualizations available within the main menu (top left-hand corner of the screen, as shown in Fig. 16.20).

Users can select the dataset containing the 4,141 health news tweets employed previously in the pipeline and press "OK." (A quick check of the dataset confirms that there are 4.1 k rows and 1 column.). In the *Data* pane, users must select "Add data source" as shown in Fig. 16.21 and select "CATEGORIES_DATA."

The *Objects* tag, under the *Data* tag, provides a list of all of the possible visualizations available on SAS VTA, as well as a list of additional analytical tools and controls. If users are interested in visualizing the popularity of each of the three categories, they can choose among multiple options. An interesting option to visualize this popularity (as measured by the total number of documents ascribed to each category) is to select *Treemap* from the *Objects* tag. Once *Treemap* is selected, VTA requires the user to specify the required roles of the data item. This task can be implemented by clicking on the *Roles* tag on the right. Under *Tile*, we selected "_category_," and under *Color*, we selected "Frequency Percent" (as shown in Fig. 16.22).

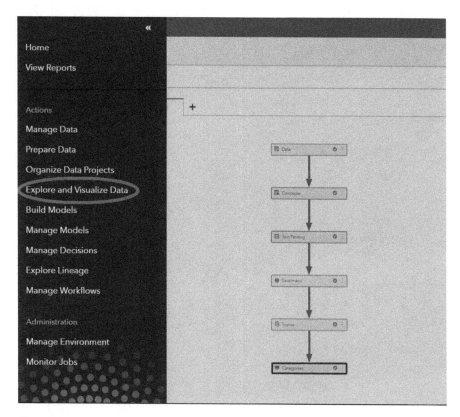

Fig. 16.20 *Explore and Visualize Data* selection for advanced analysis: main menu

These selections allow the users to visualize the results by assigning more color intensity to the most popular categories. The final visualization is shown in Fig. 16.23.

Note that the "most popular" category is obviously the "missing" category, which corresponds to those documents that cannot be classified within any of the three categories. Users can filter out these missing documents through *Filters* by removing the selection "Include missing values." Fig. 16.24 replicates the previous visualization after removing the missing values.

The results show that clearly research is the most popular category in the dataset of health news tweets from the *Los Angeles Times*, followed by disease/condition, and legal. Similar visualizations such as *Line Charts* (Fig. 16.25) or *Pie Charts* (Fig. 16.26) can be obtained through other options available through *Objects*. Frequency or frequency percentage can be employed to show the results.

As a final visualization example, users can generate a word cloud of popular terms in the dataset. To do so, the *Word Cloud* object must be selected. In this case the word cloud is implemented over key terms, not over categories. Therefore, within the *Data Roles* tag, we selected "keywords" instead of "_category_" as in the previous example. We also selected *Color* by "Frequency Percent" (see Fig. 16.27).

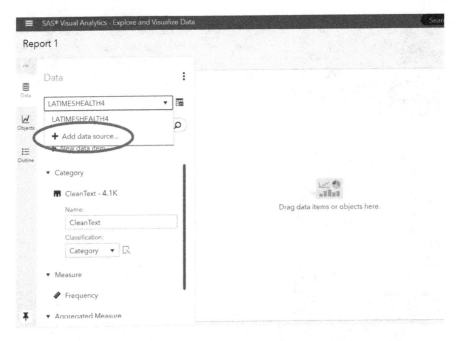

Fig. 16.21 *Add Data Source* selection

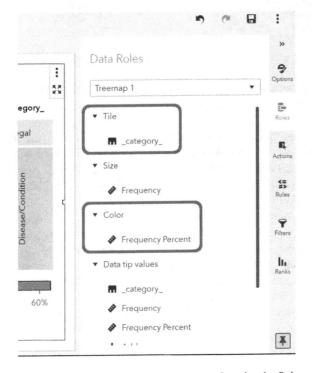

Fig. 16.22 Tile by category and color by frequency percent selected under *Roles*

Frequency, Frequency Percent by _category_

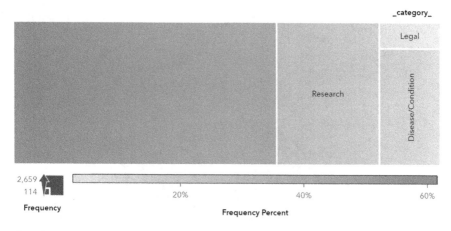

Fig. 16.23 Category popularity visualization Treemap

Frequency of _category_

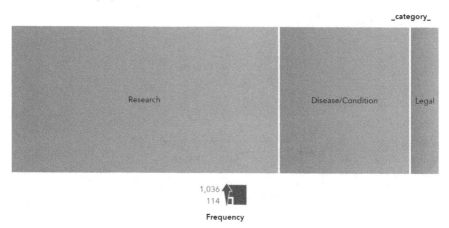

Fig. 16.24 Category popularity visualization Treemap after removing missing values

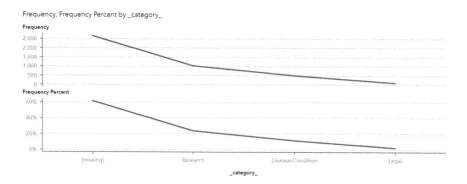

Fig. 16.25 Category popularity visualization: line charts

Frequency, Frequency Percent by _category_

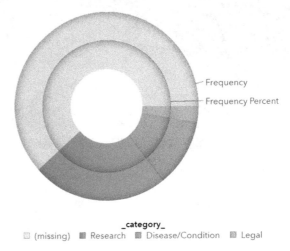

Fig. 16.26   Category popularity visualization: pie charts

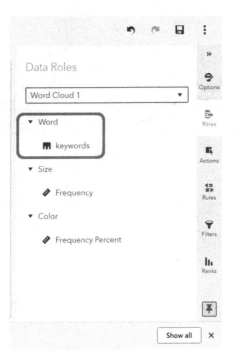

Fig. 16.27   Word by keywords and color by frequency percent selected under *Roles*

Frequency, Frequency Percent by keywords

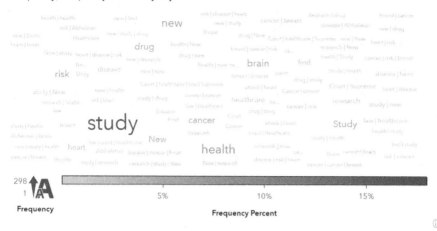

Fig. 16.28 Word cloud of key terms in the dataset

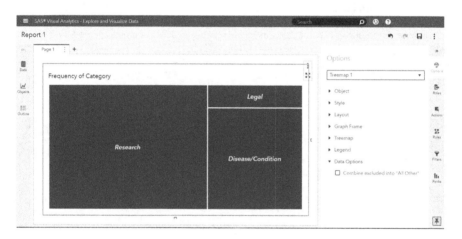

Fig. 16.29 Visualizations can be customized through the *Options* tag

The word cloud visualization is shown in Fig. 16.28.

All of the visualizations in this section were created by using the default settings on SAS VTA. Users might consider customizing those settings to develop more informative, or even more impactful, visualizations by adjusting the style, layout, or legend position through the *Options* tag located on the right-hand side of the screen, as shown in Fig. 16.29.

Key Takeaways
- SAS Visual Text Analytics provides access to advanced text analytics tools without the complexity of coding.
- SAS Visual Text Analytics performs well for frequent, repeated analyses that do not require high levels of customization.
- Data management and topic modeling are easily and robustly implemented in SAS Visual Text Analytics.

Acknowledgment The authors thank Jorge Fresneda Fernandez, Assistant Professor of Marketing at the Martin Tuchman School of Management, New Jersey Institute of Technology for contributing this chapter to the book.

Further Reading

This chapter illustrates a basic analysis implemented on SAS VTA. For more advanced options, please refer to the SAS Visual Text Analytics User's Guide (version 8.2 for this example) available at: http://documentation.sas.com/?cdcId=ctxtcdc&cdcVersion=8.2&docsetId=ctxtug&docsetTarget=titlepage.htm&locale=en

Index

CPSIA information can be obtained
at www.ICGtesting.com
Printed in the USA
LVHW082034161220
674340LV00001B/17